SPSS Manual

for

Moore and McCabe's

Introduction to the Practice of Statistics
Fourth Edition

Neal T. Rogness
Patricia A. B. Stephenson
Paul L. Stephenson
Grand Valley State University

W. H. Freeman and Company
New York

SPSS is a registered trademark of SPSS Inc.
Microsoft and Windows are registered trademarks of the Microsoft Corporation.
SPSS screen shots are reprinted with permission from SPSS Inc.
Excel 2000 and Word 2000 screen shots are reprinted with permission from the Microsoft Corporation.

ISBN 0-7167-4914-9

Copyright © 2003 by Neal T. Rogness, Patricia A. B. Stephenson, and Paul L. Stephenson

No part of this book may be reproduced by any mechanical, photographic, or electronic process, or in the form of a phonographic recording, nor may it be stored in a retrieval system, transmitted, or otherwise copied for public or private use, without written permission from the publisher.

Printed in the United States of America

First printing, 2002

Contents

Preface

0 Introduction to SPSS 11.0   1
     0.1 Navigating in SPSS 11.0   1
     0.2 Entering Data and Defining Variables   2
     0.3 Assigning Variable Value Labels   5
     0.4 Saving an SPSS Data File   7
     0.5 Opening an Existing Data File   9
     0.6 Converting an ASCII Data File to a Microsoft Excel File   9
     0.7 Saving a Microsoft Excel Data File in a Format Usable by SPSS   13
     0.8 Opening a Microsoft Excel Data File in SPSS   14
     0.9 Adding a Variable   16
     0.10 Adding an Observation   16
     0.11 Deleting a Variable   17
     0.12 Deleting an Observation   17
     0.13 Sorting Observations   17
     0.14 Selecting a Subset of Observations   18
     0.15 Recoding a Variable   19
     0.16 Printing in SPSS   21
         Printing Data in SPSS   21
         Printing Output and Charts in SPSS   21
     0.17 Deleting SPSS Output   22
     0.18 Saving SPSS Output   22
     0.19 Opening SPSS Output   23
     0.20 Copying from SPSS into Microsoft Word Office 2000   24
     0.21 Using SPSS Help   25
     0.22 Quitting SPSS   27

1 Looking at Data — Distributions   28
     1.1 Displaying Distributions with Graphs   28
         Frequency Tables, Bar Charts, and Pie Charts for Ungrouped Categorical Data   28
             Frequency Tables   28
             Bar Charts   29
             Editing Bar Charts   31
             Pie Charts   33
             Editing Pie Charts   35
         Bar Charts and Pie Charts for Grouped Categorical Data   36
             Bar Charts   37
             Pie Charts   38
         Stemplots   40
         Histograms   42
             Editing Histograms   43
         Time Plots   46
     1.2 Describing Distributions with Numbers   48
         Descriptive Statistics and Boxplots for a Single Quantitative Variable   48
         Comparing Distributions   51
         Changing the Unit of Measurement (Linear Transformations)   54
     1.3 Normal Distributions   55
         Normal Distribution Probability Calculations   55
         Normal Quantile Plots   56
         Random Number Generator   58

2 Looking at Data — Relationships 59
  2.1 Scatterplots, Correlation, and Least Squares Regression   59
  2.2 Residuals   63

3 Producing Data 67

4 Probability Distributions 70
  4.1 Binomial Probability Distributions   70
  4.2 Normal Probability Distributions   72
  4.3 Other Probability Distributions   72

5 Sampling Distributions 73
  5.1 Generating Binomial Data   73
  5.2 Generating Normal Data   74

6 Introduction to Inference 77
  6.1 Confidence Intervals   77
  6.2 Tests of Significance   78

7 Inference for Distributions 80
  7.1 Inference for the Mean of a Population   80
     One-Sample t Confidence Intervals   80
     One-Sample t Test   81
     Matched Pairs t Procedures   83
     Inference for Nonnormal Populations Using the Sign Test   86
  7.2 Two-Sample t Procedures   87

8 Inference for Proportions 90

9 Inference for Two-Way Tables 91

10 Inference for Regression 100

11 Multiple Regression 106

12 One-Way Analysis of Variance 110

13 Two-Way Analysis of Variance 116

14 Nonparametric Tests 121
  14.1 Wilcoxon Rank Sum Test   121
  14.2 Wilcoxon Signed Rank Test   123
  14.3 Kruskal-Wallis Test   126

15 Logistic Regression 128

Preface

We have found from personal experience that an ideal way to teach and learn statistics is by engaging students as active learners through exploratory data analysis. Exploratory data analysis is greatly facilitated by using statistical software because it allows students to primarily focus on the interpretation of statistical analyses, not the calculations. SPSS has been regarded as one of the most powerful statistical packages for many years. It performs a wide variety of statistical techniques ranging from descriptive statistics to complex multivariate procedures. In addition, a number of improvements have been made to version 11.0 of SPSS that make it more user-friendly. Since most of the features in SPSS 11.0 are mouse driven (requiring no programming), we think students will consider SPSS 11.0 to be an easy-to-use statistical software package.

This manual is a supplement to the fourth edition of Moore and McCabe's *Introduction to the Practice of Statistics* (IPS). The purpose of this manual is to show students how to perform the statistical procedures discussed in IPS using SPSS 11.0. This manual provides applications and examples for each chapter of the text. The statistical analyses for each example are motivated, demonstrated, and briefly described. You will observe that (for the purpose of consistency with IPS) most of the examples and subsequent discussion come directly from IPS. Step-by-step instructions describing how to carry out statistical analyses using SPSS 11.0 are provided.

Neal Rogness
Patricia Stephenson
Paul Stephenson

Chapter 0. Introduction to SPSS 11.0

This manual is a supplement to *Introduction to the Practice of Statistics*, Fourth Edition, by David S. Moore and George P. McCabe, which is referred to as IPS throughout the manual. This manual is not meant to be an exhaustive reference for SPSS 11.0, as this version abounds in both options and features. Instead, we concentrate on the techniques and procedures found in IPS, and we demonstrate how you can generate much of the same output using SPSS 11.0.

Throughout this manual, the following conventions are used: (1) variable names are given in boldface italics (e.g., ***age***), (2) commands one clicks or text one types are boldface (e.g., click **Statistics**), (3) important statistical terms are boldface, (4) the names of boxes or areas within an SPSS window are in double quotes (e.g., the "Variables Name" box), and (5), in an example number, the digit(s) before the decimal place is the chapter number and the digit(s) after the decimal place is the example number within that chapter (e.g., Example 1.3 is the third example in Chapter 1). Unless otherwise specified, all example, table, and figure numbers refer to examples, tables, and figures within this manual. If an example comes from IPS, it will be referenced as IPS Ex. c.n where c = the chapter number and n = the example number within that chapter.

This chapter serves as a brief overview of tasks that help one get started in SPSS, such as navigating in SPSS, working with SPSS data files, printing output, and using SPSS Help.

Section 0.1. Navigating in SPSS 11.0

Check with your instructor regarding any special instructions about (1) running Windows 95/98/NT at your site, and (2) where to locate and how to access SPSS on your network. Figure 0.1 shows the opening screen in SPSS 11.0 after the software has been activated. Note that in the box below the option of "Open an existing file" located within the "SPSS for Windows" window, the most recent files used in SPSS are listed, and they can be easily opened by clicking on the desired file name and then clicking "OK". If, instead, you want to enter data into a new file, you can click "Type in data" and then click "OK". For new SPSS users, we recommend clicking on "Run the tutorial" to become acquainted with the software package.

The window in the background of Figure 0.1 is the SPSS Data Editor. Note the two tabs located in the bottom left-hand corner. One tab is labeled "Data View" and the other tab is labeled "Variable View". The Data Editor is comprised of these two windows. When you open the Data Editor, by default you are in "Data View". You must be in "Data View" in order to enter data. Note that the "Data View" of Data Editor is in a spreadsheet format where columns represent variables and rows represent individual cases or observations. The SPSS Menu bar (**File**, **Edit**, ... , **Help**) appears directly below "Untitled – SPSS Data Editor". Each of these main menu options contains further submenus of additional options. Throughout this manual, the first step in performing a particular task or analysis often begins with one of the commands from the SPSS Menu bar (e.g., Click **File**, then click **Open**). The Variable View window will be discussed in Section 0.2, "Entering Data and Defining Variables".

In addition to the Data Editor, the other primary window is the Output window (which is not accessible until after output have been generated). To move between these two windows, select **Window** from the SPSS Main Menu and then click on the name of the desired window. In Figure 0.2, the SPSS Data Editor is the current active window, as shown by the ✔ in front of that window name. Another way in which to navigate between these two primary windows is to click the "Output" button at the bottom of your monitor screen to go to the Output window (this button will not appear until after output have been generated). Similarly, you can click on the button at the bottom of your monitor screen with the name of the data file to move to the Data Editor window. If the data file has not been named, the button will read "Untitled – SPSS Data Editor" (see Figure 0.3).

Most of the output in SPSS can be generated by clicking on a series of commands using a sequence of pulldown menus. However, it is also possible to generate output by writing a program using SPSS syntax language. The explanation of the syntax language is beyond the scope of this manual. However, the interested reader can learn more about this feature by clicking on **Help** in the SPSS Menu bar, then clicking on **Syntax Guide**. Programs using the SPSS syntax language are presented in Chapters 5 and 6 of this manual.

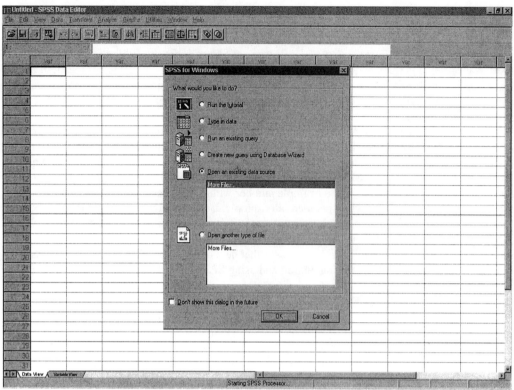
Figure 0.1

Figure 0.2

Figure 0.3

Section 0.2. Entering Data and Defining Variables

There are three primary ways in which you can create an SPSS data set: (1) you can enter data by hand (which will be explained in this section), (2) you can read in an existing SPSS data file (see Section 0.5), or (3) you can read in a data file in a different format, the common being an ASCII text file (see Section 0.6) or a Microsoft Excel file (see Section 0.8).

Prior to entering data into SPSS, you need to determine whether a variable is quantitative or categorical. A **quantitative** variable assumes numerical values for which arithmetic operations make sense, and a **categorical** variable places an individual into one of several groups or categories. SPSS also uses the variable classifications of nominal, ordinal, or scale. A **nominal** variable is a variable that classifies characteristics about the objects under study into categories. Some examples of nominal variables are eye color, race, and gender. An **ordinal** variable is a variable that classifies characteristics about the objects under study into categories that can be logically ordered. Some examples of ordinal variables are the size of eggs (small, medium, or large) and class standing (freshman, sophomore, junior, or senior). **Scale** variables, collectively referring to both interval and ratio variables, are quantitative variables for which arithmetic operations make sense. Some examples of scale variables are height, weight, and age.

For illustrative purposes, we will use the data found in Table 0.1 which is from Table 2.15, Exercise 2.105, in IPS. Note that there are three variables: ***Species***, ***Weight***, and ***Lifespan***. ***Species*** is a qualitative (and nominal) variable, whereas both ***Weight*** and ***Lifespan*** are quantitative (and scale) variables. SPSS refers to quantitative variables as "numeric" variables, and it refers to non-numeric variables (such as ***Species***) as "string" variables.

If you've not already done so, open SPSS 11.0 to the Data Editor. The first step in entering data into the Data Editor is to define the three variables of interest. In the lower left-hand corner of the Data Editor, click on the tab labeled "Variable View". A portion of the screen you should now see is in Figure 0.4. Like Data View, the Variable View sheet of the Data Editor has a spreadsheet design. Important distinctions, however, are that there are a finite number of columns on this sheet and each column is pre-labeled: Name, Type, Width, Decimals, Label, Values, Missing, Columns, Align, and Measure. We will discuss each of these columns, some in more depth than others.

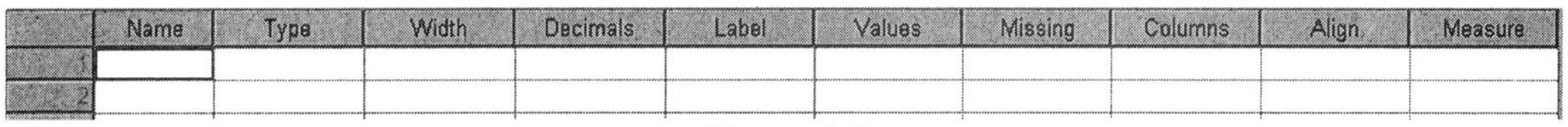

	Name	Type	Width	Decimals	Label	Values	Missing	Columns	Align	Measure
1										
2										

Figure 0.4

Table 0.1 Body weight and lifetime for several species of mammals

Species	Weight (kg)	Lifespan (years)
Baboon	32	20
Beaver	25	5
Cat, domestic	2.5	15
Chimpanzee	45	20
Dog	8.5	12
Elephant	2800	35
Goat, domestic	30	8
Gorilla	140	20
Grizzly bear	250	25
Guinea pig	1	4
Hippopotamus	1400	41
Horse	480	20
Lion	180	15
Mouse, house	0.024	3
Pig, domestic	190	10
Red fox	6	7
Sheep, domestic	30	12

We will first define the variable ***Species***. In the first cell under the "Name" column, type in **Species** and hit "Enter". Variable names in SPSS can be at most 8 characters long and cannot contain embedded blanks. Note that SPSS assigns default values under most columns in this row. By default, SPSS assumes that the variable is quantitative or numeric (Type). It also assumes that the largest data point will not exceed 8 characters (Width), that you want two decimal places of accuracy (Decimals), that there are no values to assign (Values), that there are no missing values (Missing), that you want to display 8 columns (Columns), that you want all data values to be aligned to the right side of cells (Align), and that the variable is a scale variable, since it's numeric (Measure).

The variable ***Species***, however, does not contain numeric values. Rather, it contains text (or is a "string" variable). To be able to enter the data values for the variable ***Species***, we need to change the variable type. Using the arrow keys or your mouse, move the cursor to the first cell under the "Type" column. A small gray button with three dots appears on the right side of this cell (see Figure 0.5). Click on this button and the dialog box shown in Figure 0.6 appears. Click on the circle preceding the word "String". The dialog box changes slightly (see Figure 0.7). By default, SPSS assumes that the maximum number of characters in a data value will be 8. However, the longest species name is 15 characters (Sheep, domestic). Click inside the box labeled "Characters:", and change the value of 8 to **15**. Click **OK**.

Figure 0.5

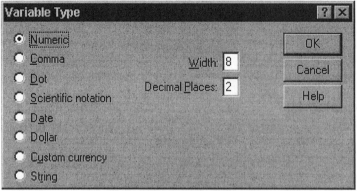
Figure 0.6

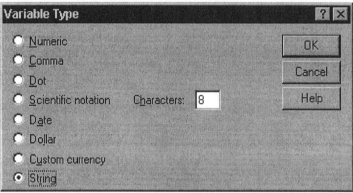
Figure 0.7

You are returned to the Variable View screen. Observe that "Type" now reads "string", that "Measure" now reads "nominal", that the width has changed from 8 to 15, and that the number of decimal points has now changed from 2 to 0. The change in "Width" will allow you to enter data values up to 15 characters in length. However, you will also want to change the column width for this variable so that all 15 characters will be <u>displayed</u>. Move the cursor to the first cell under the "Column" heading. On the right side of the cell, two small, stacked buttons appear (see Figure 0.8). The top button has an upward pointing arrow and the lower button has a downward pointing arrow. Since we want to increase the width of the column for this variable, click on the top button until the value displayed reads "15".

Figure 0.8

Since there are no missing values for this variable, we will accept the default of "none" under "Missing". Further, the concept of variable value labels does not apply here (see Section 0.3 for how to create value labels), so we will again accept the default of "None" under "Values". Also, the default of right-side alignment within a cell seems reasonable for this variable. That leaves the column of "Label". This column allows us to provide a more descriptive label for the variable, one that will appear on output associated with this variable. Recall that the maximum length of a variable name in SPSS is eight 8 characters. Often, this forces one to be creative in assigning a variable name. For instance, suppose a variable of interest was the number of children born for a household. The variable name might be "numchild" or "children", neither of which is very indicative at face value of what the variable represents. Being able to include a descriptive label such as "Number of children born" for this variable would make any output associated with this variable much more user-friendly. Variable labels can be at most 256 characters long and can contain embedded blanks. For the variable *Species*, let us use the label of "Species of animal". Click on the first cell under "Label" and type in **Species of animal**. Note that the width of this column adjusts for the length of the label.

We will now define the variable *Weight*. Click on the cell under the "Name" column that begins the second row. Type in **Weight**. This variable is numeric and scale so we can accept the default values under "Type" and "Measure". The largest data value does not exceed 8 characters, so we can accept the default values of 8 under both "Width" and "Columns". Since there are no missing values and the concept of variable values does not apply again, we can accept the default values of "None" under both "Values" and "Missing". For the variable label, type in **Weight of animal (in kg)**. As to the number of decimal values, note that the weight of a mouse is 0.024 kg. Since this involves three decimal places of accuracy, we need to change the default value of 2 under "Decimals" to 3. Move the cursor to the second cell under the "Decimals" column. You will see two small, stacked buttons on the right of the cell (like in Figure 0.8). Since we want to increase the number of decimal points from 2 to 3, click on the top button once.

Move the cursor to the third cell under the "Name" column and type in the variable name **Lifespan**. Again, we can accept the default values for the "Type", "Width", "Values", "Missing", "Columns", "Align", and "Measure" columns. For the variable label, type in **Lifespan of animal (in years)**. Also, since this variable has integer values, the number of decimal places can be reduced from 2 to 0.

All three variables associated with this problem have been appropriately defined; we can now proceed with entering the data. To enter data, you must be in "Data View" of the Data Editor, so click on the Data View tab. Using the arrow keys or your mouse, move the cursor so it is in the first cell under the variable *species*. Type in **Baboon** and hit either the "Enter" key or the down arrow (↓) key (**NOTE**: If you did not define this variable as "string", SPSS will not allow you to enter text). You will see that periods appear in the cells under both *weight* and *lifespan*, indicating that these values are currently missing. Type in **Beaver** and again hit either the "Enter" key or the down arrow key. Continue until all 17 species names have been entered. If you make a data entry mistake, just return to the cell with the incorrect value/text and reenter the data.

Move the cursor to the first cell under *weight*. Enter in the value of **32**, the weight (in kg) of a baboon, and then hit either the "Enter" key or the down arrow key. Note that the value of "32.000" appears since you defined this variable to have three decimal places of accuracy. Continue entering data for the variable *weight*, and then enter the values for the variable *lifespan* in similar fashion.

Section 0.3. Assigning Variable Value Labels

In Section 0.2, Entering Data and Defining Variables, the concept of assigning variable value labels was not discussed, mainly because the example being used did not lend itself to value labels. The purpose of this section is to provide instructions for assigning variable value labels.

Staying with the animal theme used in Example 0.1, let us say that a researcher was interested in understanding the extent to which people liked certain animals. For instance, assume that the researcher had the following five statements on a questionnaire:
   1. I like dogs.
   2. I like cats.
   3. I like horses.
   4. I like chimpanzees.
   5. I like mice.

Further, assume that a respondent is asked to provide an answer, using the following scale, for each question:
   1 = Strongly disagree
   2 = Disagree
   3 = Neutral
   4 = Agree
   5 = Strongly agree.

No doubt you have seen this type, or a similar type, of scale used before. Such a scale is often referred to as a Likert-type scale. In terms of data entry for a variable, it will be much easier to enter a number (e.g., "1") rather than having to type in the text itself (e.g., "Strongly disagree"). On the other hand, however, it would be preferential to have the text "Strongly disagree" appear on any output associated with this variable, rather than just a number. This is the situation for which having variable value labels is ideal.

If you've not already done so, activate SPSS 11.0 and go to the Variable View sheet within the Data Editor (see Section 0.1 for instructions on getting to this point). If the sheet contains information from prior work, it is best to start with a clean Data Editor. However, be sure to save your prior work if it is something you will want to use again later (see Section 0.4)! To start with a new Data Editor, click **File**, then click **New**, and then click **Data**.

To begin, let us define the first variable (see Section 0.2). Use the name "Dogs" and the variable label "I like dogs". It is reasonable to accept the default type "numeric" (since we will be entering values of 1, 2, etc.). However, under "Measure", change the measure from "scale" to "ordinal". We will also accept the default values for "Width", "Missing", "Columns", and "Align". Since the values we will enter are integers, we can change "Decimals" from 2 to 0. We are now ready to assign the variable value labels.

Figure 0.9

1. Click on the cell associated with the "Dogs" row that is under the "Values" column. A small gray box with three dots appears on the right side of the cell (see Figure 0.9).
2. Click on this button. The "Value Labels" dialog box shown in Figure 0.10 appears.

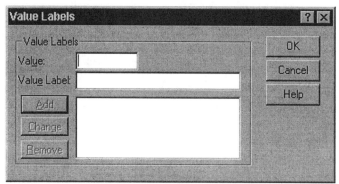
Figure 0.10

3. In the white space across from "Value:", type **1** and then hit the "Tab" key.
4. In the white space across from "Value Label:", type **Strongly disagree**.
5. Click **Add**. Note that the cursor returns to the white space across from "Value:" and that the text '1.00 = "Strongly disagree"' appears in the larger white space.
6. In the white space across from "Value:", type **2** and then hit the "Tab" key.
7. In the white space across from "Value Label:", type **Disagree**.
8. Click **Add**.
9. In the white space across from "Value:", type **3** and then hit the "Tab" key.
10. In the white space across from "Value Label:", type **Neutral**.
11. Click **Add**.
12. In the white space across from "Value:", type **4** and then hit the "Tab" key.
13. In the white space across from "Value Label:", type **Agree**.
14. Click **Add**.
15. In the white space across from "Value:", type **5** and then hit the "Tab" key.
16. In the white space across from "Value Label:", type **Strongly agree**.
17. Click **Add**.
18. Click **OK**.

To see an outcome of your work, click the Data View tab of the Data Editor. You should see a single variable with the name "dogs". Place the cursor in the first cell under this column. Let us assume that the first respondent provided a value of 4 for this variable. Type **4** and then hit the "Enter" key (or the "Tab" key). For the next four respondents, enter in the values of **1, 5, 3,** and **2**, respectively. If you prefer to see the actual labels themselves, click **View**, and then click **Value Labels**. The labels you just assigned should now appear instead of the five numerical responses you entered.

You will see that portions of some labels are not visible. To see an entire label, click the Variable View tab and adjust the value under "Columns" to **14** for this variable. Then return to the Data View sheet.

Even though the value label text appears, one can still enter the numerical equivalent. Assume the sixth respondent gave a response of "2" for this variable. Type **2** and hit the "Enter" key. The label of "Disagree" appears instead of the number "2". To have the numbers show rather than the labels, click "View". Note that a checkmark appears in front of "Value Labels" indicating that this feature is "on". To turn the feature "off", click "Value Labels".

Now that you have assigned labels for the variable *dogs*, we want to do the same thing for the other four variables. This is easy to do by using the copy and paste functions within SPSS. Make sure that you are in "Variable View" within the Data Editor.

1. Click on the gray box with a "1" at the far left of the screen (the row associated with the variable *dogs*). This will highlight or select the entire row for the first variable.
2. Hold down the "Ctrl" ("Control") key and hit the "C" key (to copy).
3. Click on the gray box with a "2" at the far left of the screen to highlight this row.
4. Hold down the "Ctrl" key and hit the "V" key (to paste). All of the values you established for the variable *dogs* are copied to the next line. Note that the variable name for the second variable appears as "*var0002*".
5. Click on the cell containing the variable name "*var0002*" and type in *cats*.
6. Repeat steps 3, 4, and 5 for the remaining three variables using the next three rows. Name variables "*var0003*" "*var0004*", and "*var0005*", *horses*, *chimps*, and *mice*, respectively.

The ability to copy and paste the variable value labels between variables is a significant improvement that first appeared with SPSS 10.0. Prior to this version, you would have had to reenter the variable value labels individually for each variable. Imagine how long this would take you with a 100-item survey that uses this response scale!

NOTE: This same copy and paste process can also be used to copy information from a single cell to another cell. In this case, you would select the cell containing the contents you would like to copy, hold down the "Ctrl" key while you hit the "C" key, click on the cell in which you would like to past the contents, and then hold down the "Ctrl" key while you hit the "V" key.

Section 0.4. Saving an SPSS Data File

It is wise to save your work and to save it often as you are working. SPSS makes a distinction between two file types you may want to save: data and output. How to save SPSS output is explained in Section 0.18. This section explains how to save an SPSS data file.

1. Click **File** and click **Save as**. The "Save Data As" window appears (see Figure 0.11).
2. If you wish to save the file on disk, click ▼ in the "Save in" box until **3 ½ Floppy [A:]** appears, and then click on this name. If you prefer to save the file in a different location, continue to click on ▼ until the name of the desired location is found, then click on this location name.
3. By default, SPSS assigns the .sav extension to data files. If you wish to save the data in a format other than an SPSS data file, click ▼ in the "Save as type" box until the name of the desired file type appears, then click on this file type name.
4. In the "File name" box, type in the desired name of the file you are saving. The name **Table_0_2** is used in Figure 0.11.

Figure 0.11

5. Right above the "File name" box, note the text "Keeping 5 of 5 variables". There are five variables in this dataset. By default, SPSS assumes you want to keep all 5 variables. If, for some reason, you want to save only certain variables from this dataset, click on the **Variables** button. The "Save Data As: Variables" window appears (see Figure 0.12). To exclude a variable from being saved, click on the ⊠ before the variable. Click **Continue**.

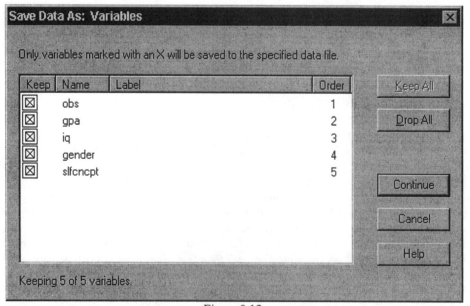

Figure 0.12

6. Make sure a disk is in the A: drive, and then click **Save**.

Section 0.5. Opening an Existing SPSS Data File

This section focuses on how to open an existing SPSS data file, either one that you previously saved, one that was given to you, or one that you obtained from another source (such as a website). In either case, we will assume that the data file is contained on a floppy disk and that the disk is in the A: drive on your computer.

If the data file is contained on a different drive, you will need to need to make the appropriate substitution in the following steps.
1. Click **File**, click **Open**, and click **Data**. The "Open File" window appears (see Figure 0.13).
2. If you wish to open an SPSS data file from a disk in the A: drive, click ▼ in the "Look in" box until **3 ½ Floppy [A:]** appears, then click on this name. If you prefer to open a file stored in a different location, continue to click on ▼ until the name of the desired location is found, then click on this location name.
3. All files with a .sav extension will be listed in the window. Click on the name of the data file you wish to open. This name now appears in the "File name" box. In Figure 0.13, **Table_0_2.sav** is the desired SPSS data file.
4. Click **Open**.

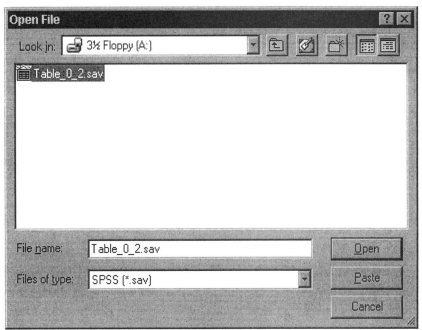

Figure 0.13

Section 0.6. Converting an ASCII Data File to a Microsoft Excel File

Although ASCII data files can be read directly into SPSS, we recommend initially reading the ASCII data file into Microsoft Excel and then reading the Microsoft Excel file into SPSS. The following example illustrates the conversion process using the data associated with IPS Table 1.6 (Educational data for 78 seventh-grade students). The data also appear in Table 0.2 of this manual. A common extension used with ASCII text data is .txt, although other extensions (i.e., .dat may represent ASCII text as well). To facilitate these steps, the data should have been entered so that the values for each variable always appear in the same columns. You can check this by looking at the data in a word processing package, such as Microsoft Word. Once the data are read in, change the font to a non-scalable font (such as Courier). If the data values for each variable do not appear in the same columns, time should be spent lining up the data before proceeding. If changes are made, be sure to save the changes to disk before continuing with these steps. For this illustration, we assume the data are contained in a data file named Table_1_6.txt which is contained on the A: drive.

Table 0.2 Educational data for 78 seventh-grade students

OBS	GPA	IQ	Gender	Self-concept	OBS	GPA	IQ	Gender	Self-concept
001	7.940	111	M	67	043	10.760	123	M	64
002	8.292	107	M	43	044	9.763	124	M	58
003	4.643	100	M	52	045	9.410	126	M	70
004	7.470	107	M	66	046	9.167	116	M	72
005	8.882	114	F	58	047	9.348	127	M	70
006	7.585	115	M	51	048	8.167	119	M	47
007	7.650	111	M	71	050	3.647	97	M	52
008	2.412	97	M	51	051	3.408	86	F	46
009	6.000	100	F	49	052	3.936	102	M	66
010	8.833	112	M	51	053	7.167	110	M	67
011	7.470	104	F	35	054	7.647	120	M	63
012	5.528	89	F	54	055	0.530	103	M	53
013	7.167	104	M	54	056	6.173	115	M	67
014	7.571	102	F	64	057	7.295	93	M	61
015	4.700	91	F	56	058	7.295	72	F	54
016	8.167	114	F	69	059	8.938	111	F	60
017	8.722	114	F	55	060	7.882	103	F	60
018	7.598	103	F	65	061	8.353	123	M	63
019	4.000	106	M	40	062	5.062	79	M	30
020	6.231	105	F	66	063	8.175	119	M	54
021	7.643	113	M	55	064	8.235	110	M	66
022	1.760	109	M	20	065	7.588	110	M	44
024	6.419	108	F	56	068	7.647	107	M	49
026	9.648	113	M	68	069	5.237	74	F	44
027	10.700	130	F	69	071	7.825	105	M	67
028	10.580	128	M	70	072	7.333	112	F	64
029	9.429	128	M	80	074	9.167	105	M	73
030	8.000	118	M	53	076	7.996	110	M	59
031	9.585	113	M	65	077	8.714	107	F	37
032	9.571	120	F	67	078	7.833	103	F	63
033	8.998	132	F	62	079	4.885	77	M	36
034	8.333	111	F	39	080	7.998	98	F	64
035	8.175	124	M	71	083	3.820	90	M	42
036	8.000	127	M	59	084	5.936	96	F	28
037	9.333	128	F	60	085	9.000	112	F	60
038	9.500	136	M	64	086	9.500	112	F	70
039	9.167	106	M	71	087	6.057	114	M	51
040	10.140	118	F	72	088	6.057	93	F	21
041	9.999	119	F	54	089	6.938	106	M	56

NOTE: This file can be downloaded as an ASCII data file from the IPS website.

To convert an ASCII text data file into a Microsoft Excel file, follow these steps.
1. Open Microsoft Excel.
2. Click **File** and then click **Open**. The "Open" window appears (see Figure 0.14).
3. Click ▼ in the "Look in" box until you find the name of the drive that contains the ASCII text data file and then click on this name. For this example, the drive that contains the ASCII text data file of interest is on drive A: (see Figure 0.14).
4. Click ▼ in "Files of type" box, and then click **All Files [*.*]**.
5. Double click **Table_1_6.txt**. The "Text Import Wizard" window opens as shown in Figure 0.15. The "Text Import Wizard" determines that the data are of fixed width (variable values are lined up in columns). As can be

seen in the "Start Import at Row" box, the "Text Import Wizard" is set to begin importing data at row 1, which is correct (since no variable names appear as column headings in the data).

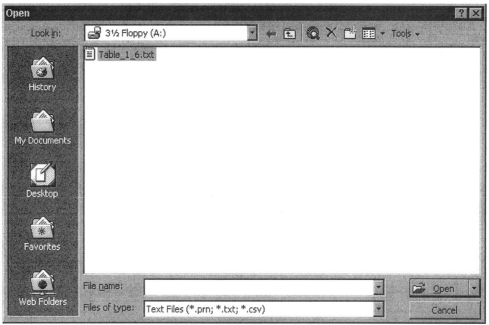

Figure 0.14

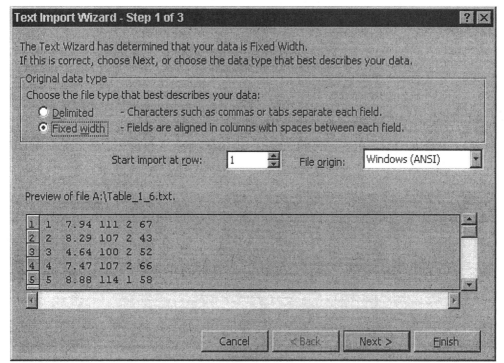

Figure 0.15

6. Click **Next>**. The second window of the "Text Import Wizard" appears, as shown in Figure 0.16. Each column of numbers in the "Data Preview" box represents the different variables in the data set. The "Text Import Wizard" suggests where the column breaks should occur between the variables. Additional column breaks can

be inserted by clicking at the desired location within the "Data Preview" box. An existing column break can be removed by double clicking on that column break. The breaks suggested by the "Text Import Wizard" are appropriate for these data.

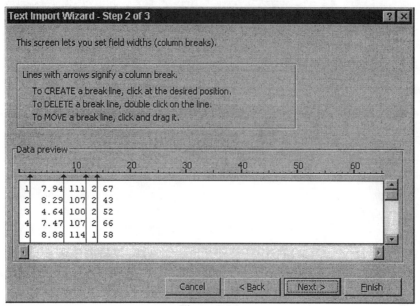

Figure 0.16

7. Click **Next>**. The third window of the "Text Import Wizard" appears, as shown in Figure 0.17. Here you can exclude variables, if desired, by clicking on the column you do not wish to import and selecting "Do Not Import Column (Skip)" in the "Column Data Format" box. You can also specify the format for each variable. By default, all variables are assigned a General format. We recommend that you retain the General format for variables to provide the most flexibility in SPSS.
8. Click **Finish**. The data will appear in the Microsoft Excel spreadsheet, where the columns represent variables and the rows represent cases. Figure 0.18 shows the values for the first five cases.

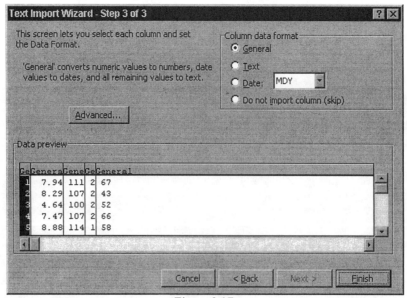

Figure 0.17

	A	B	C	D	E
1	1	7.94	111	2	67
2	2	8.29	107	2	43
3	3	4.64	100	2	52
4	4	7.47	107	2	66
5	5	8.88	114	1	58

Figure 0.18

Section 0.7. Saving a Microsoft Excel Data File in a Format Usable by SPSS

The example used in this section is based on data associated with Table 0.2 (Educational data for 78 seventh-grade students, Table 1.6 in IPS) that was introduced in Section 0.6. It is assumed that these data have already been read into Microsoft Excel (see Section 0.6) and Microsoft Excel has been opened. Before saving the Excel file, you may desire to assign the variable names in Excel. This will require that you insert a new row at the beginning of the data. You can also wait to assign the variable names when the data are read into SPSS.

To insert a new row in Excel at the beginning of data to assign variable names, follow these steps.
1. Click on the gray box with the number "1" at the far left of the screen. This will highlight or select row one.
2. Click **Insert**, and then click **Rows**. A new row will appear at the beginning of the data.

Using SPSS variable name conventions, you can now assign variable names, one name to each of the first five cells of row one. For instance, you might use the names *obs*, *gpa*, *iq*, *gender*, and *slfcncpt*, respectively (see Figure 0.19).

	A	B	C	D	E
1	obs	gpa	iq	gender	slfcncpt
2	1	7.94	111	2	67
3	2	8.29	107	2	43
4	3	4.64	100	2	52
5	4	7.47	107	2	66
6	5	8.88	114	1	58

Figure 0.19

To save a Microsoft Excel data file in a format usable by SPSS, follow these steps.
1. Click **File** and click **Save as**. The "Save As" window appears (see Figure 0.20).
2. If you wish to save the file on disk, click ▼ in the "Save in" box until **3 ½ Floppy [A:]** appears, then click on this name. If you prefer to save the file in a different location, continue to click on ▼ until the name of the desired location is found, then click on this location name.
3. You will want to save the file as a Microsoft Excel Workbook. If this option does not appear in the "Save as types:" box, click ▼ in this box until the file type **Microsoft Excel Worksheet [*.xls]** appears, then click on this file type.
4. In the "File Name" box, type in the desired name of the file you are saving. As shown in Figure 0.20, the name **Table_1_6** is used in this example.

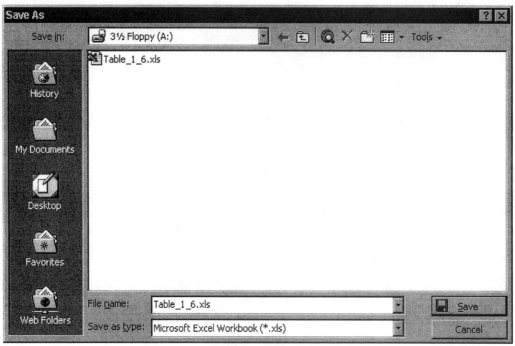
Figure 0.20

5. Make sure a disk is in the A: drive, and then click **Save**.
6. The data file will be saved on the disk in the A: drive. If you are finished with the data file in Microsoft Excel, you can close the file by clicking **File** and then **Close.** If you are finished working in Microsoft Excel, you can exit the program by clicking **File** and then **Exit**.

Section 0.8. Opening a Microsoft Excel Data File in SPSS

A common software package that many people use for data entry is Microsoft Excel. This section will provide the steps needed to read an Excel file into SPSS. For convention purposes, it is assumed that the file is contained on a floppy disk. The example used in this section is based on data found in Table 0.2, Educational data for 78 seventh-grade students (Table 1.6 in IPS). This file can be downloaded from the IPS website in an Excel format. It is assumed that the data have already been saved as a Microsoft Excel Worksheet on a floppy disk.

To open a Microsoft Excel data file in SPSS, follow these steps.
1. From the SPSS Data Editor Menu bar, click **File** and then click **Open**. The SPSS "Open File" window appears (see Figure 0.21).
2. Click ▼ in the "Look in" box until the name **3 ½ Floppy [A:]** appears, and then click on this name. If the data file was saved in a different location, continue to click on ▼ until the name of the appropriate location appears, and then click on this location name.
3. Click ▼ in the "Files of type" box until the file name type **Excel [*.xls]** appears, and then click on this file type name. The file names of any Microsoft Excel files stored on the A: drive will appear in the large center box.
4. Click on the name of the file you wish to open in SPSS. The desired file name will appear in the "File name" box. In the example, as shown in Figure 0.21, the name of the desired Microsoft Excel file is **Table_1_6.xls**.

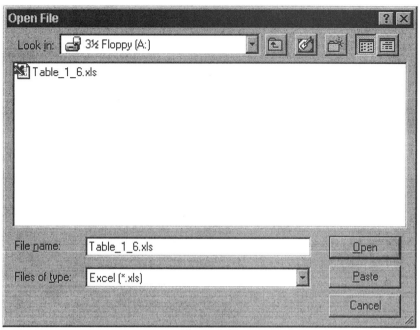

Figure 0.21

5. Click **Open**. The "Opening Excel Data Source" window appears, as shown in Figure 0.22. <u>If the dataset contains the variable names in the first row</u> (as in Figure 0.19), make sure that the box before the text "Read variable names from the first row of data" has a check mark. If it does not have a check mark, click on the box so one appears. <u>If the dataset does **not** contain the variable names in the first row</u> (as in Figure 0.18), make sure that the box before the text "Read variable names from the first row of data" does not have a check mark. If it does have a check mark, click on the box so the check mark disappears. For this example, we will assume that the variable names appear in the first row of the dataset; therefore, this box should remain checked.
6. If the Excel workbook contains multiple worksheets, you can specify which sheet SPSS should read in by clicking on the ▼ in the "Worksheet" box and clicking on the name of the desired worksheet.

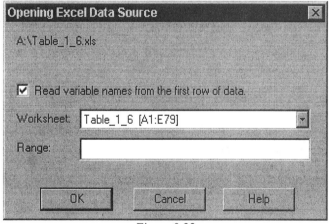

Figure 0.22

7. Click **OK**. Figure 0.23 shows the data in the SPSS Data Editor for the first five cases.

	obs	gpa	iq	gender	slfcncpt
1	1	7.94	111	2	67
2	2	8.29	107	2	43
3	3	4.64	100	2	52
4	4	7.47	107	2	66
5	5	8.88	114	1	58

Figure 0.23

Section 0.9. Adding a Variable

This section explains how to add a variable in a <u>particular</u> location. Perhaps you set up the Data Editor so that the variables are listed in the same order as they appear on a questionnaire. However, after you reach the end, you realize that you mistakenly forgot to include a variable. To facilitate data entry, you will want to have the variables appear in the Data Editor in the same order as on the questionnaire. Not to fear, this can be done with relative ease.

1. Activate the Variable View screen within the Data Editor.
2. Click on the row number (the gray box with the variable number) that corresponds to the location where you would like to insert the new variable. The entire row is highlighted. For instance, in Section 0.3, reference was made to a survey with five items. Each item represents a separate variable. The five variables were *dogs*, *cats*, *horses*, *chimps*, and *mice*. Suppose that you forgot to enter the variable *horses*, and you would like for it to now appear between the variables *cats* and *chimps*. In the Variable View screen of the Data Editor, the variable *dogs*, *cats*, *chimps*, and *mice* would appear in rows 1, 2, 3, and 4, respectively. Since you want *horses* to appear right after *cats*, click on the third row (the one currently occupied by the variable *chimps*).
3. Click **Data**, and then click **Insert Variable**. A new row appears with a generic variable name (e.g., *var0005*). The name and other defining characteristics of the new variable (*horses*) can now be entered.

   NOTE: If the placement of this overlooked variable is not important, you can simply add the variable using the next open row in the Variable View screen of the Data Editor.

Section 0.10. Adding an Observation

This section explains how to add an observation at a <u>particular</u> location. Perhaps, at the data entry stage, you overlooked a survey you received from a subject. Since each survey was assigned a unique and consecutive ID number when it was received, you want the observation to appear in its proper numerical place.

1. Activate the Data View screen within the Data Editor.
2. Click on the row number (the gray box with the observation number) that corresponds to the location where you would like to insert the new observation (case). The entire row is highlighted. For instance, assume you overlooked survey 004 and you now want it to appear as the fourth case in the dataset. You would click on the row with observation number 4.
3. Click **Data**, and then click **Insert Case**. A new row appears with a period appearing in each cell (indicating that the values are missing). The appropriate data values for each variable can now be entered.

   NOTE: If you have an ID variable included in the dataset, there is another way in which to add a case and have it appear in the appropriate numerical location. You can add the observation using the first open row at the bottom of the dataset, being sure to include its ID number. Once this is done, click **Data** and then click **Sort Cases**. Highlight the name representing the ID variable and click ▶ so this variable appears in the "Sort by:" box. Make sure that "Ascending" is selected within the "Sort Order" box. Click **OK**. The entire dataset will be resorted and the new case will appear in its appropriate numerical location. It would be a good idea to save the dataset before doing any additional work.

Section 0.11. Deleting a Variable

On occasion, you might want to remove a variable from a data set. Perhaps you created a temporary variable that is no longer needed. The steps to delete a variable are quite simple. However, you should do this with great caution, as this is not reversible!
1. Activate the Variable View screen within the Data Editor.
2. Click the row number corresponding to the variable you would like to delete (the gray box with the variable number). The entire row is highlighted.
3. Press the "Delete" key. The variable is removed from the data set.

Section 0.12. Deleting an Observation

At times, you might want to delete an entire observation from a data set. Perhaps a particular subject has a large number of missing values, or you would like to know how the analyses change if a subject is removed from the dataset. As with deleting a variable, this should be done with caution, since the process is not reversible. To recover the deleted subject(s), you will need to read back in the complete dataset. Just remember to not save the dataset that contains the deleted observation(s)!
1. Activate the Data View screen within the Data Editor.
2. Click the row number corresponding to the observation you would like to delete (the gray box with the observation number). The entire row is highlighted.
3. Press the "Delete" key. The observation is removed from the data set.

Another way in which to exclude certain observations from analyses is to create a subset of observations that does not include the observations. Advantages of this approach are that the observations are not actually removed from the data set and that the excluded observations can be added back with relative ease. Further information on this approach is provided in Section 0.14, "Selecting a Subset of Observations".

Section 0.13. Sorting Observations

You may have need to sort observations so that all observations with similar characteristics occur together in the data set. To illustrate this capability, we will use the data introduced in Section 0.6 (see Table 0.2 or Table 1.6 in IPS). Assume we want the observations to be sorted by the variable *gender* (we want all of the female observations to appear first in the dataset, followed by all of the male observations).

To sort the data, follow these steps.
1. Click **Data** and then click **Sort Cases**. The "Select Cases" window in Figure 0.24 appears.

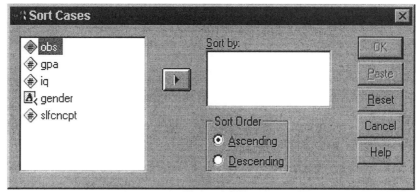

Figure 0.24

2. Click the variable *gender* and then click ▶ so *gender* appears in the "Sort by:" box. By default, SPSS assumes you want to sort in ascending order (females listed before males). If you want to sort the observations so that males are listed before females, click on the button before "Descending" in the "Sort Order" box.
3. Click **OK**. The sorted data will appear in the Data Editor.

It is possible to sort the data according to two or more characteristics. For instance, you might want to first sort the observations in ascending order according to *gender* and then to further sort the data by *gpa* by descending order within each level of *gender*. To accomplish this, follow these steps.
1. Click **Data** and then click **Sort Cases**.
2. Click on the variable *gender* and then click ▶ so *gender* appears in the "Sort by:" box. By default, SPSS assumes you want to sort in ascending order.
3. Click on the variable *gpa* and then click ▶ so *gpa* appears in the "Sort by:" box. By default, SPSS assumes you want to sort in ascending order. Click on the button before "Descending" in the "Sort Order" box.
4. Click **OK**. The sorted data will appear in the Data Editor.

NOTE: Unless you save the data when you exit SPSS, the sorted order will not be retained in the data.

Section 0.14. Selecting a Subset of Observations

You may desire to perform analyses using only a subset of observations. Perhaps you want to perform analyses using only those observations that meet a specified criterion (e.g., analyses are desired using only female respondents or you want to only use individuals who have a GPA above 2.50). To illustrate this capability, we will use the data introduced in Section 0.6 (see Table 0.2 or Table 1.6 in IPS). Assume that you only want to perform analyses using female respondents.
1. Click **Data**, and then click **Select Cases**. The "Select Cases" window in Figure 0.25 appears.

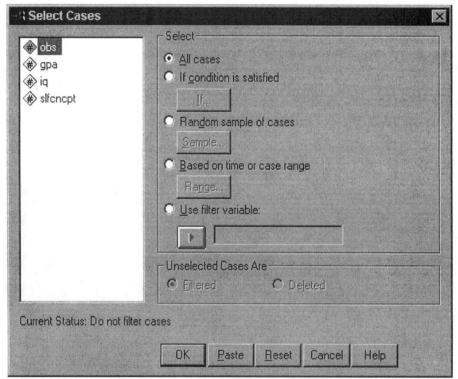

Figure 0.25

2. Click the button before "If condition is satisfied" and then click the "If" button.
3. Click *gender*, and then click ▶ to move *gender* into the upper box.

4. Click the = button and then type in **"F"** (make sure it is a capital letter, not a lowercase letter, and include the quote marks). The upper box should now appear as shown in Figure 0.26.

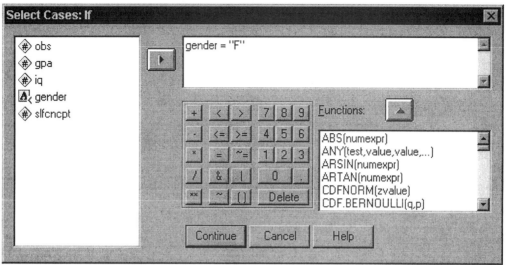

Figure 0.26

5. Click **Continue**, and then click **OK**.
6. A new variable, *filter_$*, appears in the Data Editor. All cases that have a value of "F" for *gender* have a value of 1 for *filter_$* (indicating that these cases were selected). All cases that have a value of "M" for *gender* have a value of 0 for *filter_$* (indicating these cases were not selected). Further, note that the cases with a value of "M" for *gender* have an off-diagonal line through the observation number indicating that these cases will be excluded from any subsequent analyses.

   NOTE: To include all cases in analyses, click on **Data**, click on **Select Cases**, and then click on the button before "All cases" in the "Select" section of the "Select Cases" window (see Figure 0.25). Click **OK**. The off-diagonal lines in the non-selected cases will disappear, indicating that all cases will be used in any subsequent analyses. The filter_$ variable does remain in the data set, however.

   NOTE: It is also possible to select cases based on a numerical criterion. Possible criteria might be to select cases where the gpa is strictly above 2.500 (*gpa* > **2.500**), where the iq is at or below 110 (*iq* <= **110**), or where self concept is between 60 and 80 (*slfcncpt* > **60 and** *slfcncpt* < **80**). For each example, the text that appears in parentheses is what would appear in the upper box of the "Select Cases: If" window (see Figure 0.26).

Section 0.15. Recoding a Variable

It is common to have a data file that contains both numeric and text (string) variables. However, some of the analyses performed by SPSS require that a string variable become numerical (e.g., ANOVA). If the variable has already been entered as string into the SPSS Data Editor, you can easily create a new variable that contains the same information as the string variable but is numeric simply by recoding the variable. As an example, for the data set used in Section 0.6 (see Table 0.2 or Table 1.6 in IPS), you might want to recode the numeric variable *gender* into a string variable, called *gendrtxt*, by assigning 1 = F and 2 = M.

To recode a variable into a different variable (the recommended option), follow these steps.
1. Click **Transform**, click **Recode**, and then click **Into Different Variables**. The "Recode into Different Variables" window in Figure 0.27 appears.
2. Click *gender*, and then click ▶ to move *gender* into the "Input Variable -> Output Variable" box.
3. In the "Name" box within the "Output Variable" box, type in a new variable name such as *gendrnum*.

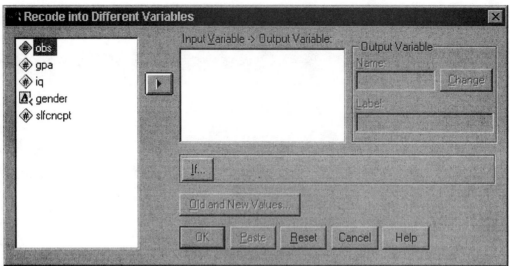

Figure 0.27

4. Click **Change**.
5. Click **Old and New Values**. The "Recode into Different Variables: Old and New Values" window in Figure 0.28 appears.
6. In the "Value" box within the "Old Value" box, type **1**. In the "Value" box in the "New Value" box, type **F**. Click **Add**. **1→'F'** appears in the "Old→ New" box. In the "Value" box within the "Old Value" box, type **2**. In the "Value" box in the "New Value" box, type **M**. Click **Add**. **2 →'M'** appears in the "Old→ New" box. Click **Continue.**
7. Click **OK**. The new variable *gendrtxt* appears in the SPSS Data Editor. To add value labels to the variable *gendrtxt*, see Section 0.3. To otherwise define this new variable, see Section 0.2.

NOTE: If you are recoding a string variable into a numeric variable, be aware that SPSS is sensitive to case. For instance, if the data includes both "F" and "f" to denote females and you want to create a new variable that recodes "F" into "1", only the observations that contain an "F" will be recoded into a "1"; the observations that contain an "f" will not be recoded.

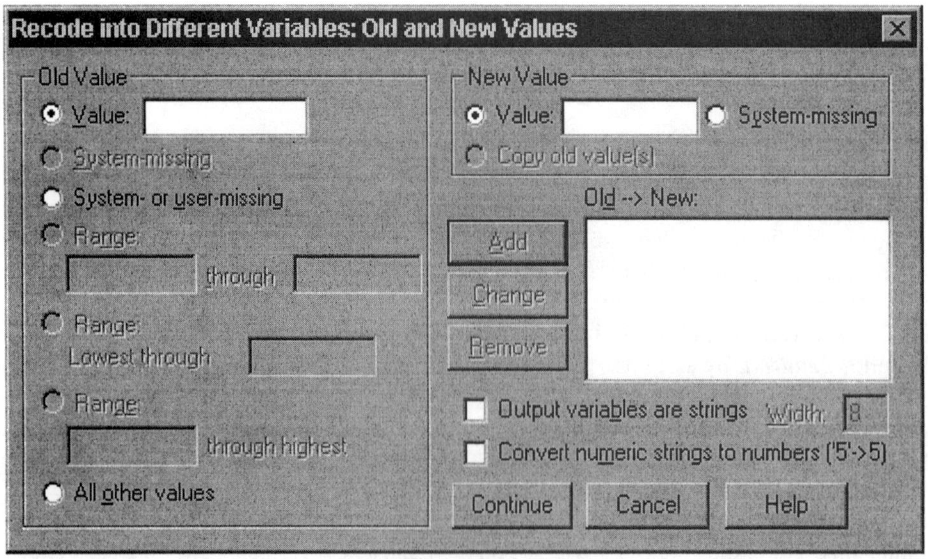

Figure 0.28

Section 0.16. Printing in SPSS

Check with your instructor regarding any special instructions about the printing process (i.e., how to select an appropriate printer). Each system tends to be unique in this manner. The remaining chapters in this manual introduce various statistical techniques, including the generation of summary statistics (called "Output") and the creation of graphs (called "Charts"). Once the desired results have been obtained, it is often of interest to print the results.

Printing Data in SPSS

To print out the data in the SPSS Data Editor, follow these steps.
1. Make the SPSS Data Editor the active window.
2. Click **File**, and then click **Print**. The "Print" window shown in Figure 0.29 appears.

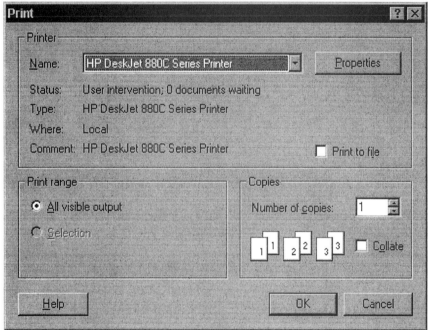

Figure 0.29

3. SPSS assumes that you want to print the entire data set ("All visible output" is selected under "Print range"). If you wish to print only a subset of the data (e.g., only one variable), you must first click on the appropriate column heading(s) in the SPSS Data Editor. If this is done, "Selection" in Figure 0.29 will be highlighted rather than "All visible output".
4. Click **OK**.

Printing Output and Charts in SPSS

To print out SPSS output and/or SPSS charts, follow these steps.
1. Make the SPSS Output window the active window.
2. If you want to print **all** of the output, including graphs, click the Output icon (Output), which appears at the top of the left-hand window. All output is highlighted, as shown in Figure 0.30.

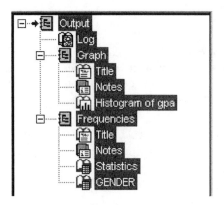

Figure 0.30

3. If you want to print only a specific portion of the output, such as the histogram of *gpa* only, click on that icon only (see Figure 0.31).

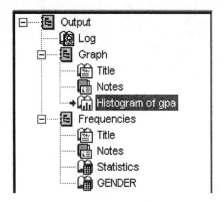

Figure 0.31

4. Click **File**, click **Print**, and then click **OK**.

Section 0.17. Deleting SPSS Output

At times, you may desire to delete either all or part of the output that you have generated. This section provides instructions for doing this. However, you only should do this if you are certain that the output will no longer be needed, as this action is not reversible. Any output mistakenly deleted will need to be regenerated.

1. Make the "SPSS Output" window the active window.
2. If you want to delete **all** of the output, click the Output icon (Output), which appears at the top of the left-hand window. All output is highlighted, as shown in Figure 0.30.
3. If you want to delete only a specific portion of the output, such as the histogram of *gpa* only, click on that icon only (see Figure 0.31).
4. Press the "Delete" key. The desired output will be deleted.

Section 0.18. Saving SPSS Output

Just as it is possible to save an SPSS data file (see Section 0.4), it is also possible to save output you have generated in SPSS. You may want to save the output you generate so that it can be read back later into SPSS for further refinement (see Section 0.19). To save SPSS output, follow these steps (it is assumed that you saved the output on a floppy disk in the A: drive).

1. Make sure that the SPSS Output Viewer is the active window.
2. If you want to save **all** of the output, including graphs, click the Output icon (Output), which appears at the top of the left-hand window. All output is highlighted, as shown in Figure 0.30. If you want to save only a specific portion of the output, such as the histogram of *gpa* only, click that icon only (see Figure 0.31).
3. Click **File** and click **Save as**. The "Save Data As" window appears (see Figure 0.32).
4. If you wish to save the file on disk, click ▼ in the "Save in" box until **3 ½ Floppy [A:]** appears, and then click on this name. If you prefer to save the file in a different location, continue to click on ▼ until the name of the desired location is found, then click on this location name.
5. By default, SPSS assigns the .spo extension to output files.
6. In the "File name" box, type in the desired name of the file you are saving. The name **Table_0_2** is used in Figure 0.32.
7. Make sure a disk is in the A: drive, and then click **Save**.

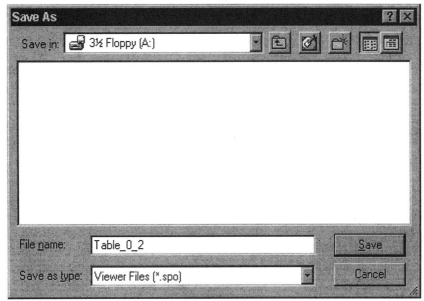

Figure 0.32

Section 0.19. Opening SPSS Output

To read in previously saved SPSS output, follow these steps (it is assumed that you saved the output on a floppy disk in the A: drive).

1. Make sure that the SPSS Data Editor is the active window.
2. Click **File**, click **Open**, and click **Output**. The "Open File" window appears (see Figure 0.33).
3. If you wish to open an SPSS data file from a disk in the A: drive, click ▼ in the "Look in" box until **3 ½ Floppy [A:]** appears, then click on this name. If you prefer to open a file stored in a different location, continue to click on ▼ until the name of the desired location is found, then click on this location name.
4. All files with a .spo extension will be listed in the window. Click on the name of the data file you wish to open. This name now appears in the "File name" box. In Figure 0.33, **Table_0_2.spo** is the desired SPSS data file.
5. Click **Open**.

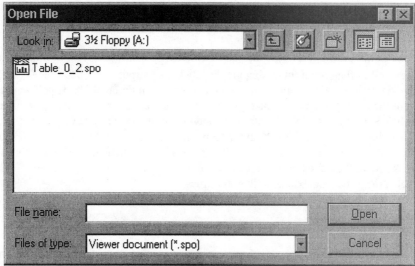
Figure 0.33

Section 0.20. Copying from SPSS into Microsoft Word Office 2000

The instructions for copying an SPSS object into Microsoft Word Office 2000 differ depending on whether the object is a chart or a table. To copy a **chart** from SPSS into Microsoft Word Office 2000, follow these steps. (Note: It is assumed that Microsoft Word Office 2000 has already been opened.)
1. In the "Output SPSS Viewer" window, select the chart to be copied by clicking on the icon that appears in the left-hand side of the window. For example, the **Histogram of gpa** icon shown in Figure 0.31.
2. Click **Edit** and then click **Copy**.
3. In Microsoft Word, position the cursor at the desired place in the document, click **Edit** and then click **Paste** (do **NOT** select Paste Special).
4. If you are interested in resizing the picture, click on the picture and place the cursor on one of the little squares appearing at the corners of the picture. Using the right mouse button, drag the selected square in or out, or up or down, depending on how you want to resize the picture.
5. To begin typing in Microsoft Word, click on any area outside the chart within the document. If you add or delete text prior to the graph, the position of the graph will adjust accordingly.

To copy a **table** from SPSS into Microsoft Word, follow these steps.
1. In the "Output SPSS Viewer" window, select the table to be copied by clicking on the icon that appears in the left-hand side of the window. For example, the **GENDER** icon (a frequency table for the variable *gender*) shown in Figure 0.31.
2. Click **Edit** and then click **Copy**.
3. In Microsoft Word, position the cursor at the desired place in the document, click **Edit** and then click **Paste Special** (do **NOT** select Paste). The "Paste Special" window in Figure 0.34 appears.
4. Click **Picture** (even though this is a table).
5. Click **OK**. The table will appear in the Microsoft Word document.
6. Click anywhere on the table. It will be outlined by eight black squares.
7. Click **Format**, **Picture**, and then the **Layout** tab.
8. Click on the icon above "Inline with text" under "Wrapping style".
9. Click **OK**.
10. To begin typing in Microsoft Word, click on any area outside the table within the document. As you add or delete text prior to the table, the position of the table will adjust accordingly.

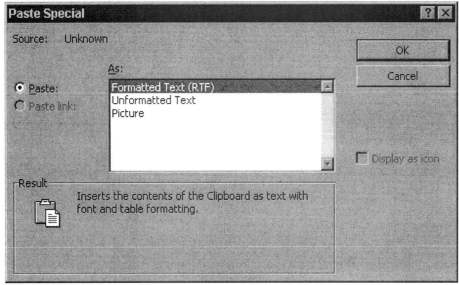

Figure 0.34

NOTE: If you click on an object (graph or table) within Microsoft Word and a box with eight solid black squares outlines the object, the object will float with text. This means that the position of the object will adjust if text (or other objects) is added prior to where the object appears. In other words, the object will retain its relative position within the document. If you click on an object and it is not outlined by a box with eight solid black squares (rather, it is surrounded by eight clear boxes), the object will not float with text (it will not retain its relative position in the document).

Section 0.21. Using SPSS Help

SPSS has extensive and useful on-line help. Suppose you want to know the steps needed to obtain a boxplot using SPSS and no manual is available. This information can be obtained using the help function of SPSS by following these steps.

1. Click **Help**, and then click **Topics**. The "Help Topics" window appears as shown in Figure 0.35.
2. Click **Index** and type **box** in the "Type the first few letters of the word you're looking for" box. Figure 0.36 shows the list of potential terms from which to choose.
3. Double click **obtaining** under "boxplots". The directions entitled "To Obtain Simple and Clustered Boxplots" appear in the "How To" window (see Figure 0.37).
4. To print the directions, click **Print**.
5. To exit SPSS Help, click ⊠ in the upper right corner of the "How To" window.

Figure 0.35

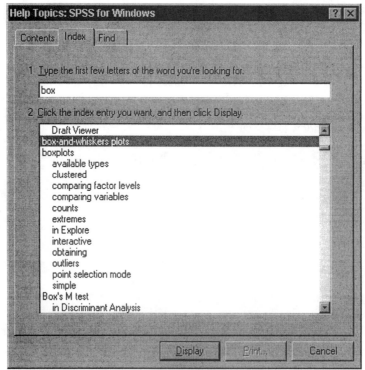

Figure 0.36

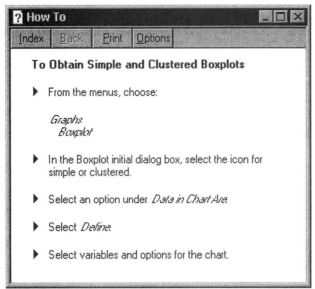

Figure 0.37

Section 0.22. Quitting SPSS

When you are finished with your work for the current session, remember to exit the program properly. To exit SPSS, follow these steps.
1. Click **File**.
2. Click **Exit**.

    If you have not saved your data and/or output since making any changes, you will be prompted as to whether you want to save the data file (see Section 0.3) or the output (see Section 0.18).

Chapter 1. Looking at Data — Distributions

Section 1.1. Displaying Distributions with Graphs

Statistical tools and ideas help us examine data. This examination is called exploratory data analysis. This section introduces the notion of using graphical displays to perform exploratory data analysis. The graphical display used to summarize a single variable will depend upon the type of variable being studied (i.e., whether the variable is categorical or quantitative). For categorical variables, **frequency tables**, **bar charts**, and **pie charts** will be examined. For quantitative variables, **stemplots**, **histograms**, and **time plots** will be examined.

Frequency Tables, Bar Charts, and Pie Charts for Ungrouped Categorical Data

The notion of obtaining a frequency table, a bar chart, and a pie chart for a small data set containing a single categorical variable is introduced here. The frequency table reports the frequencies and percentages for each of the categories of the categorical variable. The bar chart and pie chart are appropriate graphical displays used to describe a single categorical variable.

Example 1.1   Twenty randomly selected faculty members were asked if they would vote in favor of the new General Education requirements proposed to be implemented within the next two years. The data are given in Table 1.1.

Table 1.1 Votes on General Education Requirements Proposal

Yes	Yes	No	Undecided	No
Yes	No	Yes	Undecided	Yes
Yes	No	Yes	Undecided	No
No	No	Yes	Undecided	Undecided

Summarize the data set using appropriate descriptive statistics and appropriate graphs. The SPSS Data Editor contains a single variable called *vote*, which is declared type string of length 9 (see Section 0.2).

Frequency Tables

To create a frequency table for a categorical variable, follow these steps.
1. Click **Analyze**, click **Summarize**, and then click **Frequencies**. The SPSS window in Figure 1.1 appears.
2. Click *vote*, then click ▶ to move *vote* into the "Variable(s)" box.
3. Click **OK**.

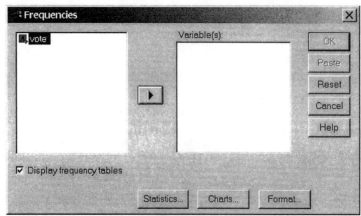

Figure 1.1

Table 1.2 is part of the resulting SPSS output.

VOTE

		Frequency	Percent	Valid Percent	Cumulative Percent
Valid	No	7	35.0	35.0	35.0
	Undecided	5	25.0	25.0	60.0
	Yes	8	40.0	40.0	100.0
	Total	20	100.0	100.0	

Table 1.2

Example 1.1 (cont.) Seven faculty members, or 35% of those sampled, would not vote in favor of the new General Education requirements. Five faculty members, or 25% of those sampled, are undecided about how they would vote. Eight faculty members, or 40% of those sampled, would vote in favor of the new General Education requirements. The Cumulative Percent is the percent of the current category plus the percents of all the categories above it.

Bar Charts

To create a bar chart for a categorical variable, follow these steps.
1. Click **Graphs** and then click **Bar**. The SPSS window in Figure 1.2 appears.
2. Click **Define**. The SPSS window in Figure 1.3 appears.
3. Click *vote*, then click ▶ to move *vote* into the "Category Axis" box.

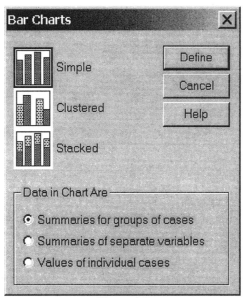

Figure 1.2

4. By default, the bars represent the number of cases. If you are interested in the *y* axis being labeled as "Percent" rather than "Count", click **% of cases** in the "Bars Represent" box.
5. If you are interested in including a title or a footnote on the chart, click **Titles**. The SPSS window in Figure 1.4 appears. In the properly labeled box ("Title" or "Footnote"), type in the desired information. Click **Continue**.
6. Click **OK**.

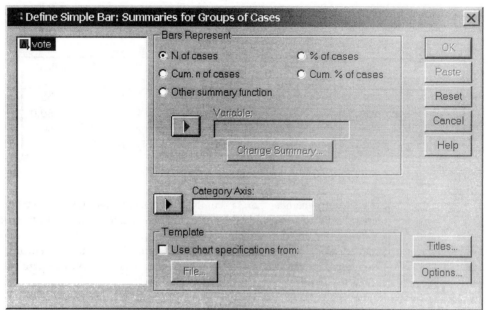

Figure 1.3

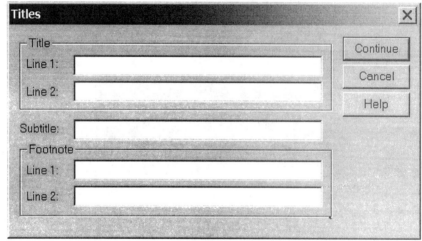

Figure 1.4

Figure 1.5 is the resulting SPSS output.

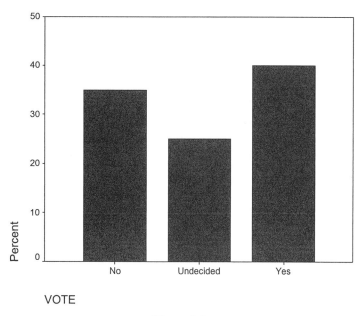

Figure 1.5

Editing Bar Charts

To have the numbers appear within the bars, follow these steps.
1. Double click on the bar chart in the "Output 1 – SPSS Viewer" window. The bar chart now appears in the "Chart 1 – SPSS Chart Editor" window, which has new menu and tool bars.
2. Click [icon]. The SPSS window in Figure 1.6 appears.

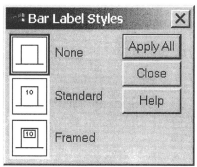

Figure 1.6

3. Click on the box in front of the **Framed** option. "None" is the default option.
4. Click **Apply All**.
5. Click **Close**.
6. If you are finished editing the chart, click **File** and then click **Close** to return to the "Output 1 – SPSS Viewer" window.

To change the color of an area within the chart, follow these steps.
1. The chart needs to be in the "Chart 1 – SPSS Chart Editor" window.
2. Click on the area within the chart for which a change in color is desired, for instance, the bars in the bar chart.

3. Click ![icon]. The SPSS window in Figure 1.7 appears.

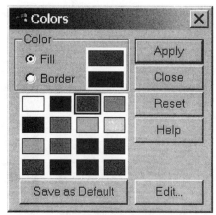

Figure 1.7

4. Make sure that "Fill" rather than "Border" is selected within the "Color" box. Click on the desired color for the fill, for instance, light gray.
5. Click **Apply**.
6. Click **Close**.
7. If you are finished editing the chart, click **File** and then click **Close** to return to the "Output 1 – SPSS Viewer" window.

To change the pattern of an area within the chart, follow these steps.
1. The chart needs to be in the "Chart 1 – SPSS Chart Editor" window.
2. Click on the area within the chart for which a change in pattern is desired, for instance, the bars in the bar chart.
3. Click ![icon]. The SPSS window in Figure 1.8 appears.

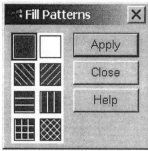

Figure 1.8

4. Click the desired pattern.
5. Click **Apply**.
6. Click **Close**.
7. If you are finished editing the chart, click **File** and then click **Close** to return to the "Output 1 – SPSS Viewer" window.

Figure 1.9 is the resulting SPSS output after adding the numbers inside the bars and changing the color of the bars to light gray.

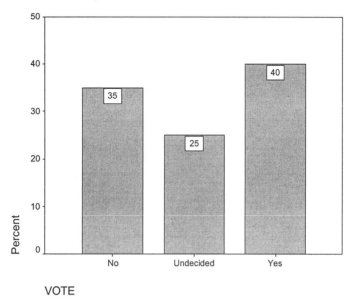

Figure 1.9

Pie Charts

To create a pie chart for a categorical variable, follow these steps.
1. Click **Graphs** and then click **Pie**. The SPSS window in Figure 1.10 appears.

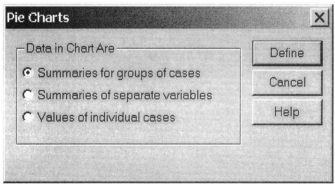

Figure 1.10

2. Click **Define**. The SPSS window in Figure 1.11 appears.
3. Click *vote*, then click ▶ to move *vote* into the "Define Slices by" box.
4. By default, the slices represent the number of cases. If you are interested in the slices representing the percents rather than the counts, click **% of cases** in the "Slices Represent" box. However, the same pie chart will appear.
5. If you are interested in including a title or a footnote on the chart, click **Titles**. The "Titles" window shown in Figure 1.4 appears. In the properly labeled box ("Title" or "Footnote"), type in the desired information. Click **Continue**.
6. Click **OK**.

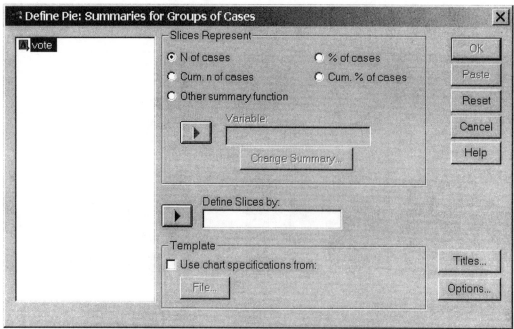

Figure 1.11

Figure 1.12 is the resulting SPSS output (except for the difference in the color scheme), after having selected **N of cases** rather than **% of cases** in the "Slices Represent" box. To change the color of a pie slice, follow the directions about changing the color in the Editing Bar Charts section.

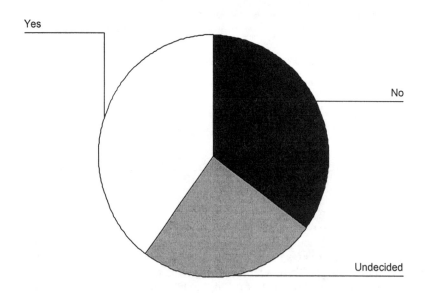

Figure 1.12

Editing Pie Charts

By default, only the category labels of the categorical variable appear on the pie chart. To add the counts and percentages to the pie slices, follow these steps.
1. Double click on the pie chart in the "Output 1 – SPSS Viewer" window. The pie chart now appears in the "Chart 1 – SPSS Chart Editor" window, which has new menu and tool bars.
2. Click **Chart** (from the new menu bar) and then click **Options**. The SPSS window in Figure 1.13 appears.
3. In the "Labels" box, click **Values** and then click **Percents**. The counts and percents appear on the outside of the pie chart beside the category text. If this is the desired format for the output, then skip to step 7.

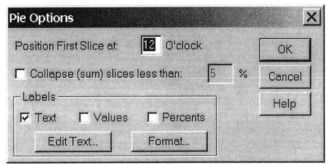

Figure 1.13

4. If you would like the numbers to appear inside the slices, then click **Format**. The SPSS window in Figure 1.14 appears.

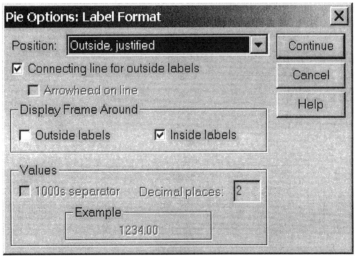

Figure 1.14

5. Click ▼ located next to the "Position" box until **Inside** is highlighted.
6. Click **Continue**.
7. Click **OK**.
8. If you are finished editing the chart, click **File** and then click **Close** to return to the "Output 1 – SPSS Viewer" window.

Figure 1.15 is the resulting SPSS output after adding the counts and percents inside the pie slices and changing the colors of the slices. To change the color or the pattern fill of a pie slice, follow the directions about changing the color or the fill in the Editing Bar Charts section.

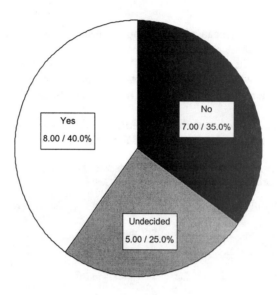

Figure 1.15

Bar Charts and Pie Charts for Grouped Categorical Data

This section introduces the notion of obtaining bar charts and pie charts for large data sets containing a single categorical variable, where the information for the single categorical variable has been entered into the SPSS Data Editor as grouped data.

Example 1.2 According to the 1996 *Statistical Abstract of the United States*, the marital distribution for all Americans age 18 and over during 1995 was the following: 43.9 million never married, 116.7 million married, 13.4 million widowed, and 17.6 million divorced. Summarize the data set using appropriate graphs. Table 1.3 shows how the data were entered into SPSS.

	marstat	count
1	nevermar	43.9
2	married	116.7
3	widowed	13.4
4	divorced	17.6

Table 1.3

The scale of measurement for *marstat* was nominal (type string 8) and for *count* was scale (type numeric 8.1). The variable *marstat* was labeled as "Marital Status" and appropriate value labels were given to each value (e.g., "nevermar" was given the value label of "Never married"). For labeling variables, follow the directions given in Section 0.3.

Bar Charts

To create a bar chart for a categorical variable that has been entered as grouped data, follow these steps.
1. Click **Graphs** and then click **Bar**. The SPSS window shown in Figure 1.2 appears.
2. Click **Values of individual cases** and then click **Define**. The SPSS window in Figure 1.16 appears.

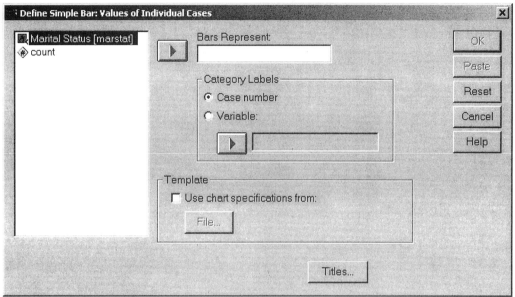

Figure 1.16

3. Click **Variable** in the "Category Labels" box. Click *marstat* and then click ▸ to move *marstat* into the "Variable" box.
4. Click *count* and then click ▸ to move *count* into the "Bars Represent" box.
5. If you want to include a title or a footnote on the chart, click **Titles**. The "Titles" window shown in Figure 1.4 appears. In the properly labeled box ("Title" or "Footnote"), type in the desired information. For this example, the following footnote was used: **Bar Graph of the Marital Status of U.S. Adults**. Click **Continue**.
6. Click **OK**.

Figure 1.17 is the resulting SPSS output. To add numbers to the bars or to change the color or the pattern fill of the bars, follow the directions given in Example 1.1 about editing bar charts.

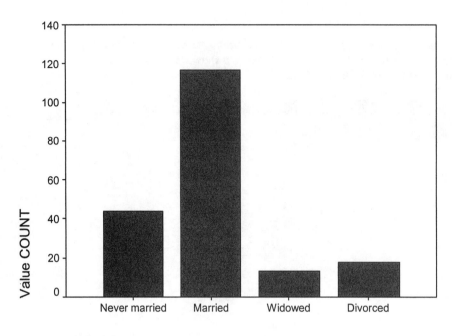

Bar Graph of the Marital Status of U.S. Adults

Figure 1.17

Pie Charts

To create a pie chart for a categorical variable that has been entered as grouped data, follow these steps.
1. Click **Graphs** and then click **Pie**. The "Pie Charts" window shown in Figure 1.10 appears.
2. Click **Values of individual cases** in the "Data in Chart Are" box and then click **Define**. The SPSS window in Figure 1.18 appears.
3. Click **Variable** in the "Slice Labels" box. Click *marstat* and then click ▶ to move *marstat* into the "Variable" box.
4. Click *count* and then click ▶ to move *count* into the "Slices Represent" box.
5. If you want to include a title or a footnote on the chart, click **Titles**. The "Titles" window shown in Figure 1.4 appears. In the properly labeled box ("Title" or "Footnote"), type in the desired information. For this example, the following footnote was used: **Pie Chart of the Same Data**. Click **Continue**.
6. Click **OK**.

    Figure 1.19 is the resulting SPSS output (except for the difference in the color scheme). To add numbers to the pie slices or change the color or the pattern fill of the pie slices, follow the directions given in Example 1.1 about editing pie charts.

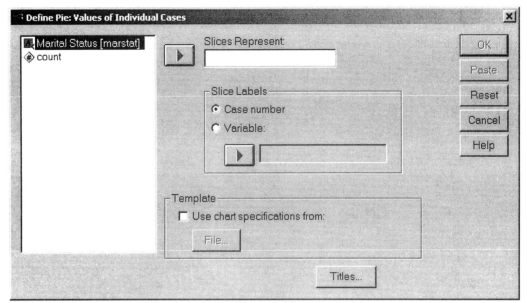

Figure 1.18

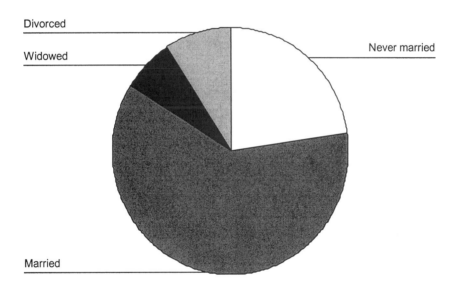

Pie Chart of the Same Data

Figure 1.19

Stemplots

A stemplot (also called a stem-and-leaf plot) gives a quick picture of the shape of the distribution for a quantitative variable while including the actual numerical values in the graph. Stemplots work best for small numbers of observations that are all greater than zero.

Example 1.3 (IPS Ex. 1.5) | A marketing consultant observed 50 consecutive shoppers at a supermarket. One variable of interest was how much each shopper spent in the store. Table 1.4 contains the data (in dollars), arranged in increasing order.

The SPSS Data Editor contains a single variable called *spending*, which is declared type numeric 8.2.

Table 1.4   Supermarket Shopping Data
3.11	8.88	9.26	10.81	12.69	13.78	15.23	15.62	17.00
17.39	18.36	18.43	19.27	19.50	19.54	20.16	20.59	22.22
23.04	24.47	24.58	25.13	26.24	26.26	27.65	28.06	28.08
28.38	32.03	34.98	36.37	38.64	39.16	41.02	42.97	44.08
44.67	45.40	46.69	48.65	50.39	52.75	54.80	59.07	61.22
70.32	82.70	85.76	86.37	93.34				

To create a stemplot of this distribution, follow these steps.
1. Click **Analyze**, click **Descriptive Statistics**, and then click **Explore**. The SPSS window in Figure 1.20 appears.

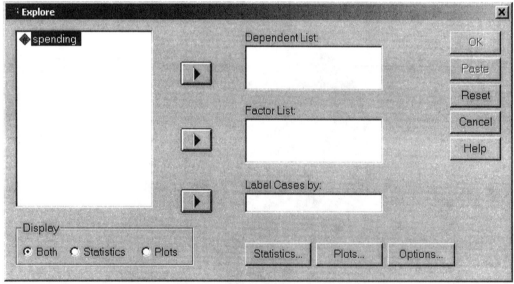

Figure 1.20

2. Click *spending*, then click ▶ to move *spending* into the "Dependent List" box.
3. By default, the "Display" box in the lower left corner has "Both" selected. Click **Plots**.
4. Click **Plots** located next to the "Options" button. The SPSS window in Figure 1.21 appears.
5. Click **None** within the "Boxplots" box. Be sure that a ✓ appears in front of "Stem-and-leaf" within the "Descriptive" box.
6. Click **Continue**.
7. Click **OK**.

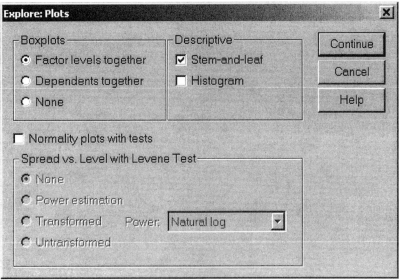

Figure 1.21

Part of the resulting SPSS output is shown in Table 1.5.

```
SPENDING Stem-and-Leaf Plot

Frequency    Stem &  Leaf
    3.00       0 .  389
   12.00       1 .  023557788999
   13.00       2 .  0023445667888
    5.00       3 .  24689
    7.00       4 .  1244568
    4.00       5 .  0249
    1.00       6 .  1
    1.00       7 .  0
    1.00       8 .  2
    3.00    Extremes    (>=86)

Stem width:    10.00
Each leaf:      1 case(s)
```

Table 1.5

Example 1.3 (cont.) The stemplot was put together by truncating the decimal places. The amount of money spent ranged from $3 to $93. The tens of dollars were used as the stems and the dollars as the leaves. The distribution of supermarket spending is skewed to the right and unimodal. According to the stemplot, there are three extreme observations above $86.

Histograms

A histogram breaks the range of values of a quantitative variable into intervals and displays only the count or the percent of the observations that fall into each interval. You can choose any convenient number of intervals, but you should always choose intervals of equal width.

Example 1.4 (IPS pp. 17–18)

The first reasonably accurate measurements of the speed of light were made over 100 years ago by A. A. Michelson and Simon Newcomb. Newcomb measured the time in seconds that a light signal took to pass from his laboratory on the Potomac River to a mirror at the base of the Washington Monument and back, a total distance of about 7400 meters. Table 1.6 contains 66 measurements made by Newcomb between July and September 1882.

Table 1.6 Newcomb's Measurements of the Passage Time of Light

28	22	36	26	28	28
26	24	32	30	27	24
33	21	36	32	31	25
24	25	28	36	27	32
34	30	25	26	26	25
– 44	23	21	30	33	29
27	29	28	22	26	27
16	31	29	36	32	28
40	19	37	23	32	29
– 2	24	25	27	24	16
29	20	28	27	39	23

Newcomb's first measurement of the passage time of light was 0.000024828 second, or 24,828 nanoseconds. There are 10^9 nanoseconds in a second. The entries in Table 1.6 record only the deviation from 24,800 nanoseconds. The entry – 44 stands for 24,756 nanoseconds. The SPSS Data Editor contains a single variable called *light*, which is declared type numeric 8.2.

To create a frequency histogram of this distribution, follow these steps.
1. Click **Graphs** and then click **Histogram**. The SPSS window in Figure 1.22 appears.
2. Click *light*, then click ▶ to move *light* to the "Variable" box.
3. If you want to include a title or a footnote on the chart, click **Titles**. The "Titles" window shown in Figure 1.4 appears. In the properly labeled box ("Title" or "Footnote"), type in the desired information. Click **Continue**.
4. Click **OK**.

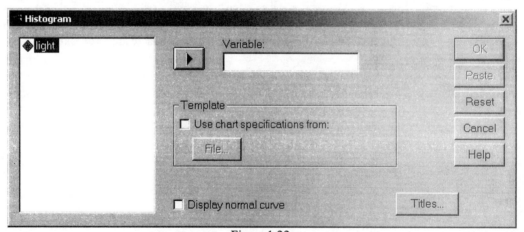

Figure 1.22

Figure 1.23 is the default histogram created by SPSS.

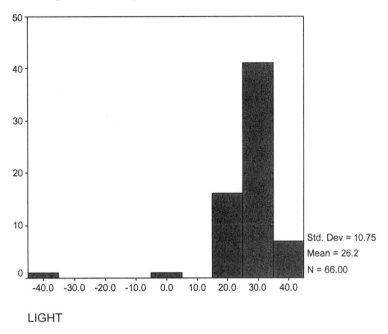

Figure 1.23

Editing Histograms

To obtain a picture of the distribution similar to that shown in IPS Figure 1.5 on page 17, the histogram can be edited. To edit the histogram, double click on the histogram in the "Output 1 – SPSS Viewer" window. The histogram now appears in the "Chart 1 – SPSS Chart Editor" window, which has new menu and tool bars.

To make changes to the x axis (such as changing the title and changing the class width of the bars), follow these steps.
1. Click **Chart** and then click **Axis**. The SPSS window in Figure 1.24 appears.

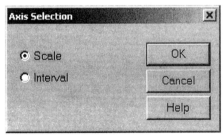

Figure 1.24

2. Click **Interval** and then click **OK**. The SPSS window in Figure 1.25 appears.
3. To change the title of the axis, replace LIGHT with the desired axis title, for instance, **Passage Time of Light**.
4. To center the axis title, click ▼ in the "Title Justification" box and click **Center**.
5. To change the range of values on the x axis and/or the class width of the bars, click **Custom** and then click **Define** located within the "Intervals" box. The SPSS window in Figure 1.26 appears.

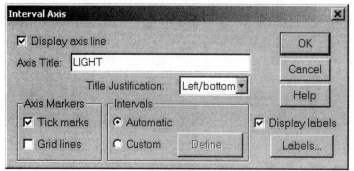

Figure 1.25

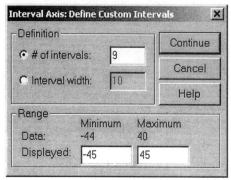

Figure 1.26

6. To change the class width, for instance to 5, click **Interval width** and then type **5** in the "Interval width" box.
7. To change the range of the *x* axis, for instance to – 60 to 60, replace – 45 with **– 60** in the "Minimum Displayed" box and replace 45 with **60** in the "Maximum Displayed" box.
8. Click **Continue**.
9. Because the class width of each bar has been decreased, there are more numbers appearing on the *x* axis, making the labels extremely difficult to read. To edit the number of labels appearing on the *x* axis, click **Labels** in the "Interval Axis" window (see Figure 1.25). The SPSS window in Figure 1.27 appears.

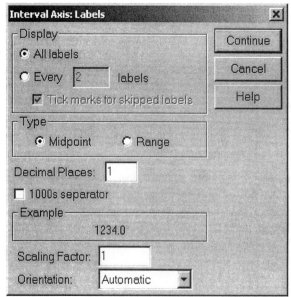

Figure 1.27

10. Click **Every ☐ labels** in the "Display" box. Replace 2 with **4** in the "Every labels" box.
11. Click **Tick marks for skipped labels**. The ✓ disappears.
12. Click ▼ in the "Orientation" box and then click **Horizontal**.
13. Click **Continue**.
14. Click **OK**.

To make changes to the *y* axis (such as adding an axis title and changing the spacing of the tick marks), follow these steps.
1. Click **Chart** and then click **Axis**. The "Axis Select" window shown in Figure 1.24 appears.
2. Click **Scale** and then click **OK**. The SPSS window in Figure 1.28 appears.
3. To label the *y* axis, type the desired label in the "Axis Title" box, such as **Frequency**.
4. To center the axis title, click ▼ in the "Title Justification" box and click **Center**.
5. To change the spacing of the tick marks on the *y* axis from 10 to 5, replace 10 with **5** in the "Major Divisions Increment" box.
6. Click **OK**.

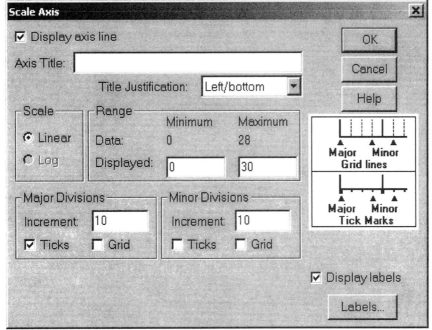

Figure 1.28

To remove the descriptive statistics (Std. Dev., Mean, and N) in the legend, follow these steps.
1. Click **Chart** and then click **Legend**. The SPSS window in Figure 1.29 appears.

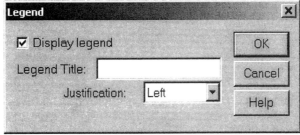

Figure 1.29

2. Click **Display legend**. The ✓ disappears.
3. Click **OK**.

Figure 1.30 is the resulting SPSS output after making changes to the *x* and *y* axes. To add numbers to the bars or change the color or the fill of the bars, follow the directions given in Example 1.1 about editing bar charts.

Example 1.4 (cont.) The histogram shows that Newcomb's data do have a symmetric unimodal distribution — but there are two low outliers (the two negative values of – 2 and – 44) that stand outside this pattern, possibly suggesting a left-skewed distribution.

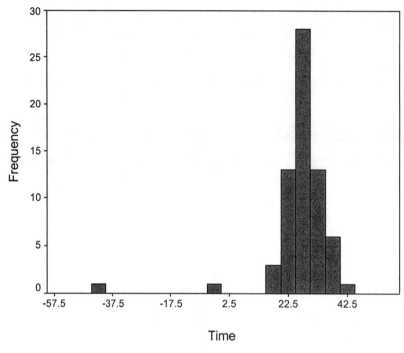

Figure 1.30

Time Plots

Many interesting data sets are time series, measurements of a variable taken at regular intervals over time. When data are collected over time, it is a good idea to plot the observations in time order. A time plot puts time on the horizontal scale of the plot and the quantitative variable of interest on the vertical scale.

Example 1.5 We can gain more insight into Newcomb's data (see Example 1.4) by examining a
(IPS pp. 18–19) time plot. Column 1 in Table 1.6 gives the first 11 measurements made by Newcomb, the second column gives the next 11 measurements made by Newcomb, etc. The time plot for this data set will plot Newcomb's passage times of light against the sequence number, where the sequence number reveals the order in which the observations were recorded.

The SPSS Data Editor contains a single variable called *light*, which is declared type numeric 8.2.

To obtain a time plot of a quantitative variable against its sequence number, follow these steps.
1. Click **Graphs** and then click **Sequence**. The SPSS window in Figure 1.31 appears.
2. Click *light*, then click ▶ to move *light* to the "Variables" box.
3. Click **OK**.

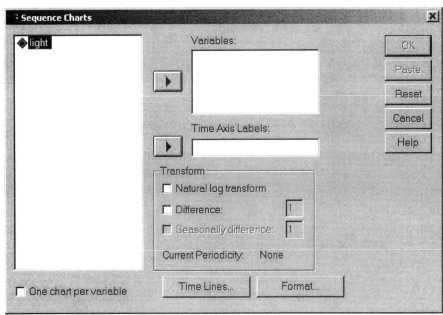

Figure 1.31

Figure 1.32 is the default SPSS output. To change the labels on the *x* axis (also called the category axis in SPSS) and the *y* axis (also called the scale axis in SPSS), consult Example 1.4 and the directions in the Editing Histograms section.

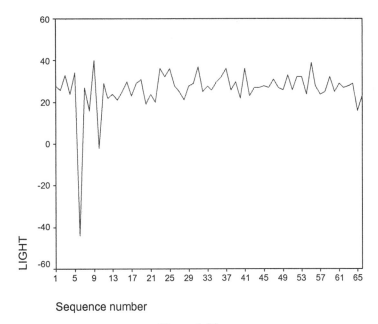

Figure 1.32

Example 1.5 (cont.)

There is some suggestion in this plot that the variability (the vertical spread in the plot) is decreasing over time. In particular, both outlying observations were made early on. Perhaps Newcomb became more adept at using his equipment as he gained experience.

Section 1.2. Describing Distributions with Numbers

This section introduces the notion of describing distributions with numbers. A description of a distribution almost always includes a **measure of center** and a **measure of spread**. To get a quick summary of both center and spread, use the five-number summary (the **minimum**, the **25th percentile**, the **median**, the **75th percentile**, and the **maximum**). The five-number summary leads to a visual representation of the distribution called a **boxplot**.

Descriptive Statistics and Boxplots for a Single Quantitative Variable

Example 1.6
(IPS Ex. 1.16)

A marketing consultant observed 50 consecutive shoppers at a supermarket. One variable of interest was how much each shopper spent in the store. See Table 1.4 for the data. Obtain descriptive statistics and a boxplot for the quantitative variable *spending*. The SPSS Data Editor contains a single variable called *spending*, which is declared type numeric 8.2.

To obtain descriptive statistics (such as the mean, median, standard deviation, percentiles, etc.) for a quantitative variable, follow these steps.
1. Click **Analyze**, click **Descriptive Statistics**, and then click **Explore**. The "Explore" window shown in Figure 1.20 appears.
2. Click *spending*, then click ▶ to move *spending* into the "Dependent List" box.
3. By default, the "Display" box in the lower left corner has **Both** selected. Click **Statistics**.
4. Click the **Statistics** button located in lower right corner of the window. The SPSS window in Figure 1.33 appears.
5. Click **Percentiles**. Be sure that a ✔ appears in front of "Descriptives".
6. Click **Continue**.
7. Click **OK**.

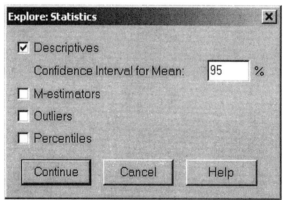

Figure 1.33

Tables 1.7 and 1.8 are the resulting SPSS output.

Percentiles

		Percentiles						
		5	10	25	50	75	90	95
Weighted Average(Definition 1)	SPENDING	9.0890	12.7990	19.0600	27.8550	45.7225	69.4100	86.0345
Tukey's Hinges	SPENDING			19.2700	27.8550	45.4000		

Table 1.7

48

Descriptives

			Statistic	Std. Error
SPENDING	Mean		34.7022	3.06848
	95% Confidence Interval for Mean	Lower Bound	28.5359	
		Upper Bound	40.8685	
	5% Trimmed Mean		33.2422	
	Median		27.8550	
	Variance		470.777	
	Std. Deviation		21.69740	
	Minimum		3.11	
	Maximum		93.34	
	Range		90.23	
	Interquartile Range		26.6625	
	Skewness		1.103	.337
	Kurtosis		.709	.662

Table 1.8

Example 1.6 (cont.) The mean and the median amount spent in the store are $34.70 and $27.86, respectively. The amount of money spent varies from $3.11 to $93.34. The standard deviation is $21.70. SPSS uses slightly different rules than IPS to compute the quartiles, so the results given by the computer may not agree exactly with the results found by using IPS rules. SPSS uses two methods (labeled "Weighted" and "Tukey's Hinges" (see Table 1.8)) to calculate the quartiles. The answers between the methods differ slightly ($19.06 versus $19.27 and $45.72 versus $45.40).

To create a boxplot for a quantitative variable, follow these steps.
1. Click **Graphs** and then click **Boxplot**. The SPSS window in Figure 1.34 appears.

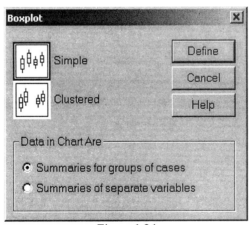

Figure 1.34

2. Click **Summaries of separate variables** and then click **Define**. The SPSS window in Figure 1.35 appears.
3. Click *spending*, then click ▸ to move *spending* into the "Boxes Represent" box.
4. Click **OK**.

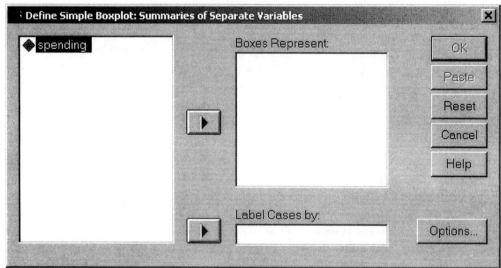

Figure 1.35

Figure 1.36 is the resulting SPSS output (except for the difference in the color scheme). To change the color or the pattern fill of the boxplot, follow the directions given in Example 1.1 about editing bar charts.

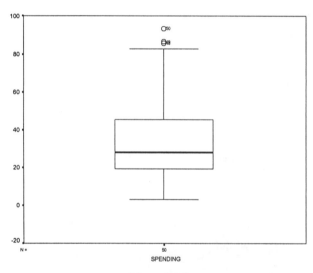

Figure 1.36

Example 1.6 (cont.)

SPSS distinguishes between minor and major outliers. Minor outliers (denoted by ° in the plot) are observations more than $1.5 \times$ IQR outside the central box. Major outliers (denoted by * in the plot) are observations more than $3.0 \times$ IQR outside the central box. SPSS also puts the observation number beside the symbol used for the outlier. For this example, there are three minor outliers occurring at observation numbers 48, 49, and 50.

Comparing Distributions

Example 1.7
(IPS Ex. 1.17)

The Environmental Protection Agency lists most vehicles into one of two categories: "minicompact" or "two-seater". We are interested in comparing the gas mileage of minicompact cars with two-seater cars and their respective city mileages with their highway mileages. Table 1.9 gives the city and highway mileage for all of the cars in these groups. We can make the comparison by using some numerical tools and side-by-side boxplots for describing distributions. The SPSS data editor contains two variables called *type* (declared type string 8) and *mileage* (declared type numeric 8.1). The variable *type* contains the following values: MiniHwy, MiniCity, TwoHwy, and TwoCity.

Table 1.9 Gas Mileage for the Groups of Cars
Minicompact Highway : 31, 27, 28, 23, 24, 22, 28, 24, 30, 25, 22
Minicompact City: 22, 19, 20, 16, 17, 16, 20, 18, 22, 17, 15
Two-seater Highway: 24, 30, 28, 27, 21, 26, 21, 16, 13, 68, 26, 13, 28, 23, 19, 27, 23, 27, 30
Two-seater City: 17, 22, 21, 20, 13, 18, 11, 11, 8, 61, 20, 10, 22, 16, 13, 21, 17, 19, 25

To obtain descriptive statistics (such as the mean, median, standard deviation, percentiles, etc.) for a quantitative variable broken down by a categorical variable, follow these steps.
1. Click **Analyze**, click **Descriptive Statistics**, and then click **Explore**. The "Explore" window shown in Figure 1.20 appears, with the appropriate variable names in the window.
2. Click *mileage*, then click ▶ to move *mileage* into the "Dependent List" box.
3. Click *type*, then click ▶ to move *type* into the "Factor List" box.
4. By default, the "Display" box in the lower left corner has **Both** selected. Click **Statistics**.
5. Click the **Statistics** button located in lower right corner of window. The "Explore Statistics" window shown in Figure 1.33 appears.
6. Click **Percentiles**. Be sure that "Descriptives" is checked.
7. Click **Continue**.
8. Click **OK**.

Tables 1.10 and 1.11 are part of the resulting SPSS output.

Percentiles

			\multicolumn{7}{c}{Percentiles}
		TYPE	5	10	25	50	75	90	95
Weighted Average(Definition 1)	MILEAGE	MiniCity	15.000	15.200	16.000	18.000	20.000	22.000	.
		MiniHwy	22.000	22.000	23.000	25.000	28.000	30.800	.
		TwoCity	8.000	10.000	13.000	18.000	21.000	25.000	.
		TwoHwy	13.000	13.000	21.000	26.000	28.000	30.000	.
Tukey's Hinges	MILEAGE	MiniCity			16.500	18.000	20.000		
		MiniHwy			23.500	25.000	28.000		
		TwoCity			13.000	18.000	21.000		
		TwoHwy			21.000	26.000	27.500		

Table 1.10

Descriptives

	TYPE			Statistic	Std. Error
MILEAGE	MiniCity	Mean		18.364	.7295
		95% Confidence Interval for Mean	Lower Bound	16.738	
			Upper Bound	19.989	
		5% Trimmed Mean		18.348	
		Median		18.000	
		Variance		5.855	
		Std. Deviation		2.4196	
		Minimum		15.0	
		Maximum		22.0	
		Range		7.0	
		Interquartile Range		4.000	
		Skewness		.307	.661
		Kurtosis		-1.163	1.279
	MiniHwy	Mean		25.818	.9517
		95% Confidence Interval for Mean	Lower Bound	23.698	
			Upper Bound	27.939	
		5% Trimmed Mean		25.742	
		Median		25.000	
		Variance		9.964	
		Std. Deviation		3.1565	
		Minimum		22.0	
		Maximum		31.0	
		Range		9.0	
		Interquartile Range		5.000	
		Skewness		.344	.661
		Kurtosis		-1.240	1.279
	TwoCity	Mean		19.211	2.5633
		95% Confidence Interval for Mean	Lower Bound	13.825	
			Upper Bound	24.596	
		5% Trimmed Mean		17.512	
		Median		18.000	
		Variance		124.842	
		Std. Deviation		11.1733	
		Minimum		8.0	
		Maximum		61.0	
		Range		53.0	
		Interquartile Range		8.000	
		Skewness		3.089	.524
		Kurtosis		11.796	1.014
	TwoHwy	Mean		25.789	2.6298
		95% Confidence Interval for Mean	Lower Bound	20.265	
			Upper Bound	31.314	
		5% Trimmed Mean		24.155	
		Median		26.000	
		Variance		131.398	
		Std. Deviation		11.4629	
		Minimum		13.0	
		Maximum		68.0	
		Range		55.0	
		Interquartile Range		7.000	
		Skewness		2.872	.524
		Kurtosis		10.997	1.014

Table 1.11

To create side-by-side boxplots for a quantitative variable broken down by a categorical variable, follow these steps.
1. Click **Graphs** and then click **Boxplot**. The "Boxplot" window shown in Figure 1.34 appears.
2. Click **Define**. The SPSS window in Figure 1.37 appears.

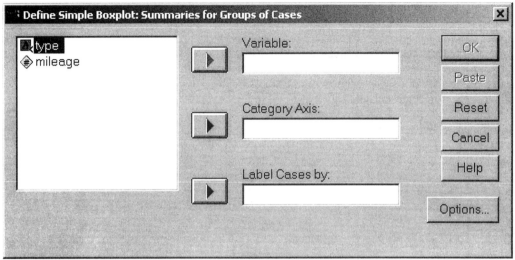

Figure 1.37

3. Click *mileage*, then click ▶ to move *mileage* into the "Variable" box.
4. Click *type*, then click ▶ to move *type* into the "Category Axis" box.
5. Click **OK**.

Figure 1.38 is the resulting SPSS output (except for the difference in the color scheme). To change the color or the pattern fill of the boxplots, follow the directions given in Example 1.1 about editing bar charts.

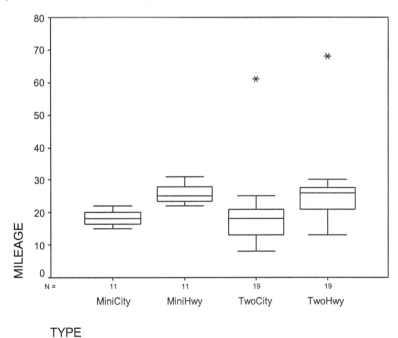

Figure 1.38

Example 1.7 (cont.) From the side-by-side boxplots, we see at once that in both minicompact cars and two-seater cars highway mileage exceeds city mileage. For both types of cars the first quartile of the highway mileage equals or exceeds the third quartile of the city mileage. One can also observe that the spread for the distributions of the mileage for the two-seater cars exceeds the spread for the minicompact cars.

Changing the Unit of Measurement (Linear Transformations)

The same variable can be recorded in different units of measurement. Americans commonly record distances in miles and temperatures in degrees Fahrenheit, while the rest of the world measures distances in kilometers and temperatures in degrees Celsius. Fortunately, it is easy to convert numerical descriptions of a distribution from one unit of measurement to another. This is true because a change in the measurement unit is a **linear transformation** of the measurements.

Example 1.8 (IPS Ex. 1.20) Suppose you have a data set containing the variable temperature measured in degrees Fahrenheit. In order for the results to be understood by the rest of the world, the variable should be reexpressed in degrees Celsius. The linear transformation is tempcel = (5/9)tempfahr – (160/9), where tempcel = the temperature expressed in degrees Celsius and tempfahr = the temperature expressed in degrees Fahrenheit. The SPSS Data Editor contains a single quantitative variable called *tempfahr*, which is declared numeric 8.2

To create a new variable in the SPSS Data Editor that is a linear transformation of an existing variable in the SPSS Data Editor, follow these steps.
1. Click **Transform** and then click **Compute**. The SPSS window in Figure 1.39 appears.

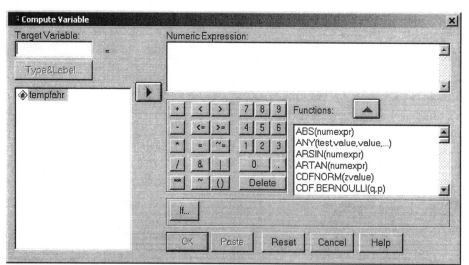

Figure 1.39

2. In the "Target Variable" box, type in *tempcel*.
3. Click *tempfahr*, then click ▶ to move *tempfahr* into the "Numeric Expression" box.
4. In the "Numeric Expression" box, type in the following expression: 5/9*tempfahr* – 160/9.
5. Click **OK**. The variable *tempcel* now appears in the "SPSS Data Editor" window.

Section 1.3. Normal Distributions

One particularly important class of density curves is the **normal distribution**. These density curves are symmetric, unimodal, and bell-shaped. All normal distributions have the same overall shape. The exact density curve for a particular normal distribution is specified by giving its mean µ and its standard deviation σ. This section discusses how to use SPSS to do normal distribution probability calculations, how to obtain **normal quantile plots**, and how to generate observations from a normal distribution.

<u>Normal Distribution Probability Calculations</u>

Example 1.9	The level of cholesterol in the blood is important because high cholesterol levels increase the risk of heart disease. The distribution of blood cholesterol levels in a large population of people of the same age and sex is roughly normal. For 14-year-old boys, the mean is 170 milligrams of cholesterol per deciliter of blood (mg/dl) and the standard deviation is 30 mg/dl. Using software, find what proportion of 14-year-old boys have less than 240 mg/dl of cholesterol. What proportion of 14-year-old boys have more than 240 mg/dl of cholesterol? What proportion of 14-year-old boys have blood cholesterol between 170 and 240 mg/dl?

To obtain the proportion of interest, follow these steps.
1. Enter the variables and values of *x1* and *x2* into the SPSS Data Editor, where *x1* = 170 and *x2* = 240.
2. Click **Transform** and then click **Compute**. The "Compute" window shown in Figure 1.39 appears with the exception that *x1* and *x2* appear in the window.
3. In the "Target Variable" box, type in *prop*.
4. In the "Functions" box, click ▼ until **CDF.NORMAL(q, mean, stddev)** appears in the box. Double click on **CDF.NORMAL(q, mean, stddev)** to move **CDF. NORMAL(?, ?, ?)** into the "Numeric Expression" box. The CDF.NORMAL(q, mean, stddev) function stands for the cumulative distribution function for the normal distribution, and it calculates the area to the left of *q* under the correct normal curve.
5. In the "Numeric Expression" box, change the second question mark to **170** and the third question mark to **30** (the appropriate values for the mean and the standard deviation).
6. For the proportion less than 240, **CDF.NORMAL(x2, 170, 30)** should appear in the "Numeric Expression" box. For the proportion more than 240, **1 − CDF.NORMAL(x2, 170, 30)** should appear in the "Numeric Expression" box. For the proportion between 170 and 240, **CDF.NORMAL(x2, 170, 30) − CDF.NORMAL(x1, 170, 30)** should appear in the "Numeric Expression" box.
7. Click **OK**.

The variable *prop* can be found in the SPSS Data Editor. By default, the number of decimal places for the variable *prop* is two. The number of decimal places can be changed; follow the directions given in Section 0.8.

Example 1.9 (cont.)	The proportion of 14-year-old boys with blood cholesterol less than 240 is 0.9902. The proportion of 14-year-old boys with blood cholesterol more than 240 is 0.0098. The proportion of 14-year-olds with blood cholesterol between 170 and 240 is 0.4902.

The previous example showed how to find the relative frequency of a given event. The next example shows how to find the observed value corresponding to a given relative frequency.

Example 1.10 (IPS Ex. 1.28)	Scores on the SAT verbal test in recent years follow approximately a normal distribution with a mean of 505 and a standard deviation of 110. Using software, determine how high a student must score to place in the top 10% of all students taking the SAT. After following the steps on the next page, you can see that a student must score at least 646 to place in the top 10% of all students taking the SAT.

To obtain the observed value corresponding to a given relative frequency, follow these steps.
1. Enter the variable and the value of *x1* into the SPSS Data Editor, where *x1* represents the area under the curve to the left of the desired score. Thus, *x1* = 1 − 0.10 = 0.90.
2. Click **Transform** and then click **Compute**. The "Compute" window, shown in Figure 1.39, appears with *x1* in the window.
3. In the "Target Variable" box, type *score*.
4. In the "Functions" box, click ▼ until **IDF.NORMAL(p, mean, stddev)** appears in the box. Double click on **IDF. NORMAL(p, mean, stddev)** to move **IDF. NORMAL(?, ?, ?)** into the "Numeric Expression" box. The IDF.NORMAL(p, mean, stddev) function stands for the inverse of the cumulative distribution function for the normal distribution, and it calculates the X value such that the area to the left of X under the correct normal curve is *p*.
5. In the "Numeric Expression" box, change the first question mark to *x1*, the second question mark to **505** and the third question mark to **110**. Thus, **IDF.NORMAL(x1, 505, 110)** should appear in the "Numeric Expression" box.
6. Click **OK**. The variable *score* can be found in the SPSS Data Editor.

<u>**Normal Quantile Plots**</u>

A useful tool for assessing normality is a graph called the normal quantile plot. Any data that follow a normal distribution produce a straight line on the normal quantile plot. Systematic deviations from a straight line indicate a nonnormal distribution. Outliers appear as points that are far away from the overall pattern of the plot.

Example 1.11 **(IPS Ex. 1.28)**	Consider Newcomb's light measurement data given in Table 1.6. Assess normality for the distribution by examining the normal quantile plot.
	The points deviate from the straight line, especially the two negative values of − 2 and − 44, suggesting a nonnormal distribution. The fact that the two lowest values of the passage time of light deviate substantially to the left of the line suggests a left-skewed distribution. This agrees with the histogram shown in Figure 1.30.

To create a normal Q-Q plot of this distribution, follow these steps.
1. Click **Analyze**, click **Descriptive Statistics**, and then click **Explore**. The "Explore" window shown in Figure 1.20 appears with the variable *light* in the window.
2. Click *light*, then click ▶ to move *light* into the "Dependent List" box.
3. By default, the "Display" box in the lower left corner has **Both** selected. Click **Plots**.
4. Click **Plots** located next to the "Options" button. The "Explore Plots" window shown in Figure 1.21 appears.
5. Click **None** within the "Boxplots" box.
6. Click **Stem-and-leaf** within the "Descriptive" box. The ✓ disappears.
7. Click **Normality plots with tests**. A ✓ appears in the box.
8. Click **Continue**.
9. Click **OK**.

Figure 1.40 is part of the resulting SPSS output.

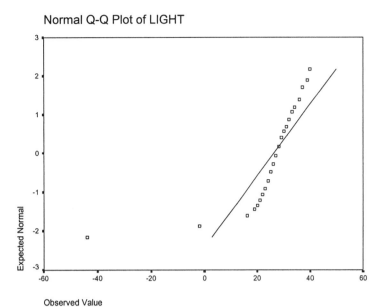

Figure 1.40

Example 1.11 (cont.) A normal Q-Q plot of Newcomb's data with the two negative values omitted is presented in Figure 1.41. The effect of omitting the outliers is to magnify the plot of the remaining data. Most of the points lie close to the straight line, indicating that a normal model fits well.

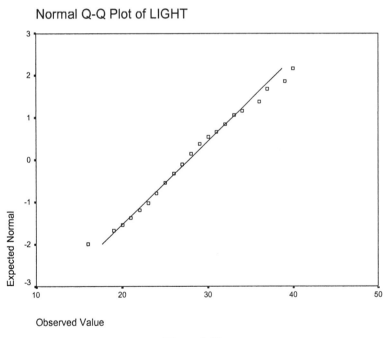

Figure 1.41

Random Number Generator

Example 1.12 — Use software to generate 100 observations from a normal distribution with a mean of 25 and a standard deviation of 5.

To generate observations from a normal distribution, follow these steps.
1. In the SPSS Data Editor, create a variable called *rannorm* that is declared type numeric 8.2 and has 100 missing values (that is, the first 100 cells under the column header *rannorm* should have periods in them).
2. Click **Transform** and then click **Compute**. The "Compute" window, as shown in Figure 1.39, appears with the appropriate variable names in the window.
3. In the "Target Variable" box, type in *rannorm*.
4. In the "Functions" box, click ▼ until **RV.NORMAL(mean, stddev)** appears in the "Functions" box. Double click on **RV. NORMAL(mean, stddev)** to move **RV. NORMAL(?, ?)** into the "Numeric Expression" box.
5. In the "Numeric Expression" box, change the first question mark to **25** and the second question mark to **5** (the appropriate values for the mean and the standard deviation).
6. Click **OK**. SPSS asks the question: "Change existing variable?" Click **OK**.

Chapter 2. Looking at Data — Relationships

Section 2.1. Scatterplots, Correlation, and Least Squares Regression

This section introduces analysis of two variables that may have a linear relationship. In analyzing two quantitative variables, it is useful to display the data in a **scatterplot**, determine the **correlation** between the data, and find the **least squares regression** line. A scatterplot is a graph that puts one variable on the *x* axis (the *explanatory* or *independent* variable) and the other on the *y* axis (the *response* or *dependent* variable), and it is used to determine whether an overall pattern exists between the variables. The correlation measures the strength and direction of the linear relationship, and the least squares regression line is the equation of the line that best represents the data.

Example 2.1 (IPS Ex. 2.10) How do children grow? The pattern of growth varies from child to child, so we can best understand the general pattern by following the average height of a number of children. Table 2.1 presents the mean heights of a group of children in Kalama, an Egyptian village that was the site of a study of nutrition in developing countries. The data were obtained by measuring the heights of 161 children from the village each month from 18 to 29 months of age.

Table 2.1 Mean Height of Kalama Children

Age	18	19	20	21	22	23	24	25	26	27	28	29
Mean Height	76.1	77.0	78.1	78.2	78.8	79.7	79.9	81.1	81.2	81.8	82.8	83.5

To generate a scatterplot, follow these steps.
1. Click **Graphs**, click **Scatter**, and click **Define**. The "Simple Scatterplot" window in Figure 2.1 appears.
2. Click *age*, then click ▶ to move *age* into the "X Axis" box.
3. Click *height*, then click ▶ to move *height* into the "Y Axis" box.
4. Click **OK**.

Example 2.1 (cont.) Figure 2.2 is the SPSS scatterplot of the data set. Age is the explanatory variable (plotted on the *x* axis), and height is the response variable (plotted on the *y* axis). The plot shows a strong, positive, linear association with no outliers. The next step is to obtain the correlation coefficient and the equation of the regression line.

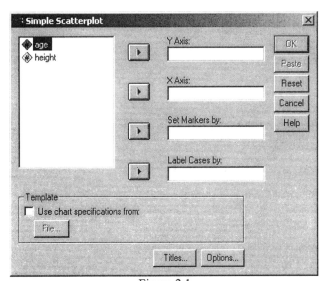

Figure 2.1

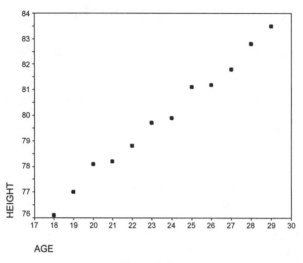

Figure 2.2

To obtain statistics such as the correlation, *y*-intercept, and slope of the regression line, follow these steps.
1. Click **Analyze**, click **Regression**, and then click **Linear**. The "Linear Regression" window in Figure 2.3 appears.
2. Click *age*, then click ▸ to move *age* into the "Independent(s)" box.
3. Click *height*, then click ▸ to move *height* into the "Dependent" box.
4. Click **OK**.

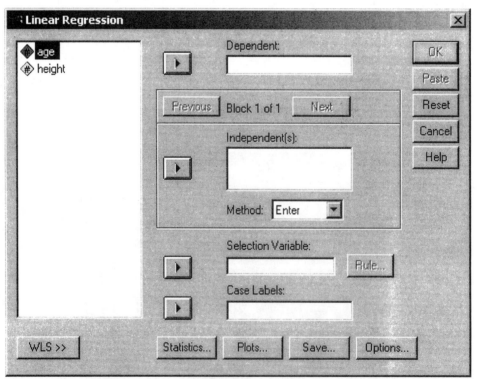

Figure 2.3

Tables 2.2 and 2.3 are part of the SPSS output that shows the correlation, the *y*-intercept, and the slope.

Model Summary

Model	R	R Square	Adjusted R Square	Std. Error of the Estimate
1	.994[a]	.989	.988	.2560

a. Predictors: (Constant), AGE

Table 2.2

Coefficients[a]

Model		Unstandardized Coefficients		Standardized Coefficients	t	Sig.
		B	Std. Error	Beta		
1	(Constant)	64.928	.508		127.709	.000
	AGE	.635	.021	.994	29.665	.000

a. Dependent Variable: HEIGHT

Table 2.3

Example 2.1 (cont.) The correlation is $r = 0.994$, and $r^2 = 0.989$, found in Table 2.2. The value of r confirms that these data have a strong, positive, linear association. The intercept is $a = 64.928$, and the slope is $b = 0.635$, found in Table 2.3 in the "Unstandardized Coefficients" column. The *y*-intercept is the value in the row labeled "(Constant)", and the slope is denoted by its variable name, in this case, ***age***. Therefore, the equation of the least squares regression line is
$$\hat{y} = 64.928 + 0.635x.$$

To plot the least squares regression line on the scatterplot, follow these steps.
1. Generate the scatterplot as described at the beginning of this chapter.
2. When the scatterplot appears in the output window, double click inside the scatterplot to gain access to the Chart Editor.
3. Once in the Chart Editor, click **Chart**, click **Options**, and then click **Total**, which is found in the "Fit Line" box (see Figure 2.4).
4. Click **Fit Options**. The window in Figure 2.5 appears.
5. The "Fit Method" defaults to "Linear regression," which is what is desired.
6. If you are interested in displaying R^2 in the legend of the scatterplot, click **Display R-square in legend**.
7. Click **Continue**.
8. Click **OK**.

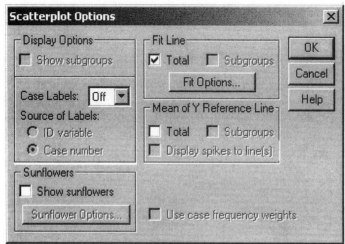

Figure 2.4

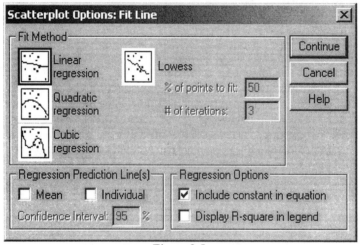

Figure 2.5

Example 2.1 (cont.) Figure 2.6 is the least squares regression line plotted on the scatterplot of the Kalama children data set. Recall that the equation of the line is $\hat{y} = 64.928 + 0.635x$, and the correlation is $r = 0.994$. Such a high correlation indicates that the linear relationship is indeed very strong, and the relationship is confirmed by the way the points cluster around the least squares regression line in Figure 2.6.

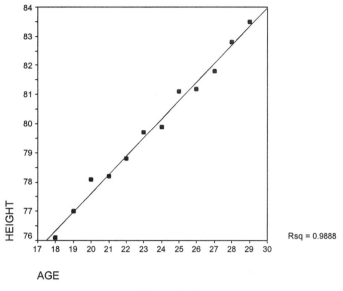

Figure 2.6

Section 2.2. Residuals

A **residual** is the difference between an observed value of the response variable and the value predicted by the regression line, written $residual = y - \hat{y}$. The primary purpose for analyzing residuals is to determine whether the linear model best represents a data set. **Residual plots** are a type of scatterplot in which the independent variable is often on the x axis and the residuals are on the y axis. This type of plot is used to detect patterns that may exist by magnifying the deviations from the line. It is desirable for no pattern to exist on the residual plot, but instead for the plot to be an unstructured band centered around $y = 0$. If this is the case, then a linear fit is appropriate. If a pattern does exist on the residual plot, it could indicate that the relationship between y and x is nonlinear or that perhaps the variation of y is not constant as x increases. Residual plots are also useful in identifying outliers and influential observations.

To generate the residuals, follow these steps.
1. Click **Analyze**, click **Regression**, and then click **Linear**.
2. Click *age*, then click ▸ to move *age* into the "Independent(s)" box.
3. Click *height*, then click ▸ to move *height* into the "Dependent" box.
4. Click **Save**, then click **Unstandardized** which is found in the "Residuals" box (see Figure 2.7).
5. Click **Continue**, then click **OK**.

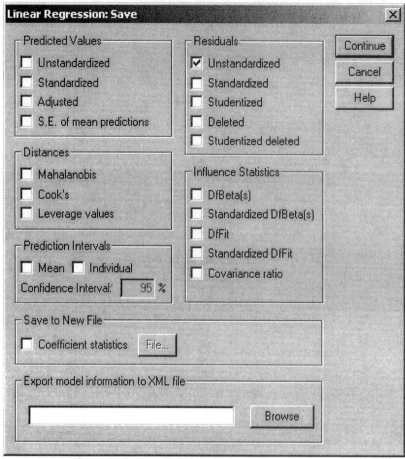

Figure 2.7

The residuals for this data set have been generated, saved, and added to the Data Editor, as shown in Figure 2.8.

To plot the residuals against the independent variable, in this case, *age*, and to plot the reference line at $y = 0$ on this plot, follow these steps.
1. Click **Graphs**, click **Scatter**, and then click **Define**.
2. Click *age*, then click ▸ to move *age* into the "X Axis" box.
3. Click *Unstandardized Residuals (res_1)*, then click ▸ to move *Unstandardized Residuals (res_1)* into the "Y Axis" box.
4. Click **OK**.
5. When the scatterplot appears in the output window, double click inside the scatterplot to gain access to the Chart Editor.
6. Click **Chart**, click **Reference Line**, and then click **Y scale** (see Figure 2.9). This places a horizontal reference line that crosses the *y* axis.
7. Click **OK**. The "Scale Axis Reference Lines" window in Figure 2.10 appears.
8. Since a horizontal line at $y = 0$ is desirable, and the field entitled "Position of Line(s)" defaults at 0, click **Add**.
9. Click **OK**. The resulting plot is in Figure 2.11.

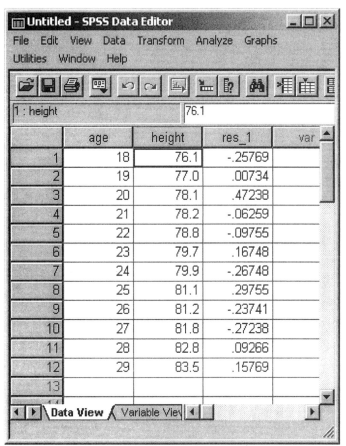

Figure 2.8

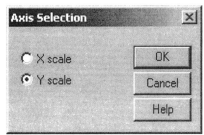

Figure 2.9

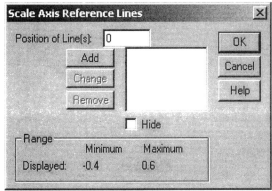

Figure 2.10

Example 2.1 (cont.) Note how the residuals of the Kalama data in Figure 2.11 are randomly scattered around the line $y = 0$. This indicates that the linear model is an appropriate model for this data set.

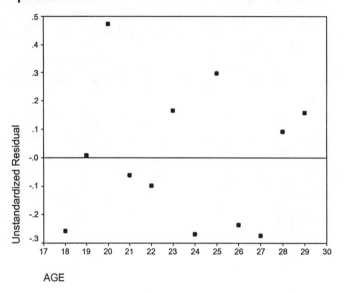

Figure 2.11

NOTE: SPSS has the capability to perform more advanced regression diagnostics, such as DFFITS (difference between fitted values), studentized residuals, and other techniques. To get to the SPSS window that produces the data necessary for such techniques, follow the instructions presented earlier for generating the residuals. Please refer to Figure 2.7, the SPSS window "Linear Regression: Save". The diagnostic statistics are in the box entitled "Influence Statistics".

Chapter 3. Producing Data

This chapter demonstrates how SPSS can be used to **select *n* cases** from a finite population of interest using **simple random sampling**. Simple random sampling, the most basic sampling design, allows impersonal chance to choose the cases for inclusion in the sample, thus eliminating bias in the selection procedure.

Example 3.1
(IPS Ex. 3.16)

An academic department wishes to choose a three-member advisory committee at random from the 28 faculty members of the department who are listed in Table 3.1.

Table 3.1 Faculty Names

00	Abbott	07	Goodwin	14	Pillotte	21	Theobald
01	Cicirelli	08	Haglund	15	Raman	22	Vader
02	Cuellar	09	Johnson	16	Riemann	23	Wang
03	Dunsmore	10	Keegan	17	Rodriguez	24	Wieczorek
04	Engle	11	Luo	18	Rowe	25	Williams
05	Fitzpatrick	12	Martinez	19	Sommers	26	Wilson
06	Garcia	13	Nguyen	20	Stone	27	Wong

One option is to use a table of random digits. Another option is to use SPSS to randomly choose 3 of the 28 faculty members. A data file was created using the variable *name* (declared as a string variable with length 12), which contained the names of the 28 faculty members of the department.

To select a random sample of *n* cases from *N* cases, follow these steps.
1. Click **Data**, and click **Select Cases**. The "Select Cases" window in Figure 3.1 appears.
2. Click **Random sample of cases** and then click the **Sample** button. The "Select Cases: Random Sample" window in Figure 3.2 appears.
3. Click **Exactly** ☐ **cases from the first** ☐ **cases** and fill in the boxes so that the line reads "Exactly **3** cases from the first **28** cases".
4. Click **Continue**.
5. Click **OK**.

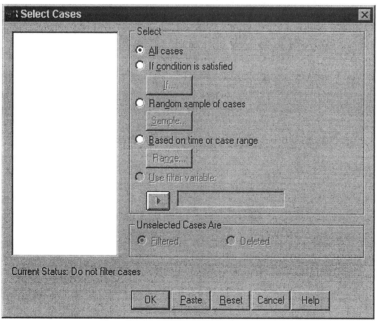

Figure 3.1

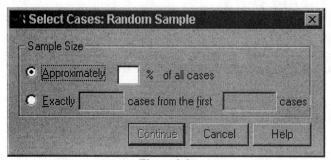

Figure 3.2

Example 3.1 (cont.)

SPSS creates a new variable, *filter_$*, which is assigned a value of 0 if the case was not randomly selected and a value of 1 if the case was randomly selected. Further, unselected cases are marked in the Data Editor with a diagonal line through the row number, as can be seen in Figure 3.3, which shows the first 10 cases. With this procedure, faculty members Abbott, Rowe, and Wieczorek were randomly selected to serve on the advisory committee. If the *filter_$* variable is deleted and the procedure repeated, a different set of faculty members should be randomly selected to serve.

	name	filter_$
1	Abbott	1
2	Cicirelli	0
3	Cuellar	0
4	Dunsmore	0
5	Engle	0
6	Fitzpatrick	0
7	Garcia	0
8	Goodwin	0
9	Haglund	0
10	Johnson	0

Figure 3.3

While this example illustrates how SPSS can be used to select a random sample of *n* cases from a finite population, this same process can be used to randomly assign subjects to treatments. The names listed in Table 3.1 represent a finite population of interest for Example 3.1. However, for purposes of illustration, assume that the individuals listed in Table 3.1 instead represent a <u>random sample</u> of faculty from a larger population of interest. Further, suppose that the researcher wants to investigate if faculty learn a new teaching activity better by attending a hands-on workshop or by teaching themselves using a self-paced, interactive computer module.

Once the sample has been selected, SPSS can be used to randomly determine who attends the workshop (Treatment A) and who learns the activity via the computer (Treatment B). With the cases of two treatments, the exact same steps presented earlier in this chapter can be followed. The only exception would be that one would select "Exactly **14** cases from the first **27** cases". The key here is to obtain a random subset of all cases from the sample. **NOTE:** Before testing this out, delete the existing *filter_$* variable by clicking on the column heading for this variable and pressing the "Delete" key. Since there is an odd number of faculty, assume that it was randomly determined that 14 faculty would attend the workshop and 13 faculty would learn via the computer. As before, SPSS will create a new *filter_$* variable in which all observations are assigned either a 1 (selected) or a 0 (unselected). The 13 selected faculty will attend the workshop and the 14 unselected faculty will utilize the computer.

If there are three treatment levels instead of two, SPSS can be used to randomly assign subjects to treatments, but slightly more work is needed to accomplish this goal. Again, for illustrative purposes, assume that the faculty names in Table 3.1 represent a random sample from a larger population of interest. Further, assume that the researcher wants to assess how well faculty learn a new teaching activity while comparing Treatment A (attending a hands-on workshop), Treatment B (doing a self-paced, interactive computer module), and Treatment C (reading printed materials that explain the activity). Since there are 27 faculty names, the researcher wants to randomly assign nine faculty to each of the three treatments. To start, follow the same steps presented earlier in this chapter with the following exception: select "Exactly **9** cases from the first **27** cases." **NOTE:** Before testing this out, delete the existing *filter_$* variable by clicking on the column heading for this variable and pressing the "Delete" key. Once again, SPSS will create a new *filter_$* variable in which all observations are assigned either a 1 (selected) or a 0 (unselected). The nine selected faculty will be assigned Treatment A. To determine which faculty will be assigned Treatments B and C, continue with the following additional steps.

1. Click **Data** and click **Select Cases**.
2. On the "Select Cases" window (see Figure 3.1), select **All Cases**.
3. Click **OK**.

NOTE: The *filter_$* variable will remain, but the diagonal lines in the case number columns will disappear.

4. Click **Data** and click **Sort Cases**.
5. Click the *filter_$* variable and click ▶ to move *filter_$* into the "Sort by:" box.
6. Click **Ascending** within the "Sort Order" box.
7. Click **OK**.

NOTE: The 18 unselected faculty (those with a value of 0 under the *filter_$* variable) are now all listed first followed by the nine selected faculty (those with a value of 1 under the *filter_$* variable).

8. Click **Data** and click **Select Cases**.
9. On the "Select Cases" window (see Figure 3.1), select **Random sample of cases** and then click the **Sample** button.
10. On the "Select Cases: Random Sample" window (see Figure 3.2), click **Exactly ☐ cases from the first ☐ cases** and fill in the boxes so that the line reads "Exactly **9** cases from the first **18** cases".
11. Click **Continue**.
12. Click **OK**.

SPSS will assign the values of 1 and 0 within the *filter_$* variable so that, within the first 18 cases, half of the names will have a value of 1 and half will have a value of 0 (the last nine names, the nine faculty previously selected to receive Treatment A, also now have a value of 0). Among the first 18 faculty names listed, those with a value of 1 under the *filter_$* variable will receive Treatment B and those with a value of 0 under the *filter_$* variable will receive Treatment C.

Chapter 4. Probability Distributions

This chapter will show how SPSS can be used to find **probabilities** associated with both **binomial** and **normal probability distributions**. In addition, a list of other probability distributions for which SPSS can calculate probabilities is given at the end of the chapter.

Section 4.1. Binomial Probability Distributions

Example 4.1 | A quality control engineer selects a simple random sample of 10 switches from a large
(IPS Ex. 5.4) | shipment for detailed inspection. Unknown to the engineer, 10% of the switches in the shipment fail to meet the specifications. What is the probability that no more than 1 of the 10 switches in the sample fails inspection?

Let X = the number of switches in the sample that fail to meet the specifications. X is a binomial random variable with $n = 10$ and $p = 0.10$. To find the probability that no more than 1 of the 10 switches in the sample fails inspection ($P(X \leq 1 \mid n = 10, p = 0.10)$), follow these steps.

1. Define a new variable *switch*, which takes on the values 0 through 10. This variable is declared a numeric variable of length 8.0.
2. Click **Transform** and then click **Compute**. The "Compute Variable" window in Figure 4.1 appears.

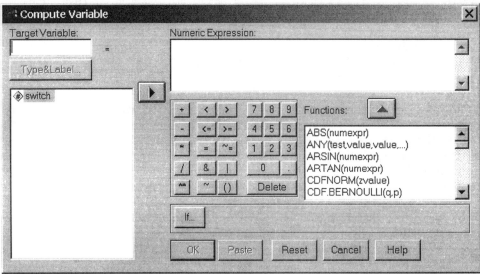

Figure 4.1

3. In the "Target Variable" box, type in *lesseq*.
4. In the "Functions" box, click ▼ until **CDF.BINOM(q,n,p)** appears in the box. Double click **CDF.BINOM(q,n,p)** to move CDF.BINOM(?,?,?) into the "Numeric Expression" box. The CDF.BINOM(q,n,p) function stands for the cumulative distribution function for the binomial distribution, and it calculates the cumulative probability that the variable takes a value less than or equal to q.
5. In the "Numeric Expression" box, highlight the first *?*, click *switch*, and click ▶ so that *switch* replaces the first *?*. Delete the second *?* and replace it with the number **10** (the value of *n*), and delete the third *?* and replace it with the value **0.10** (the value of *p*).
6. Click **OK**.

Figure 4.2 displays the new variable *lesseq*. By default, SPSS shows the values of *lesseq* to two decimal places of accuracy. The number of decimal places was changed to four (see Section 0.2).

70

switch	lesseq
0	.3487
1	.7361
2	.9298
3	.9872
4	.9984
5	.9999
6	1.0000
7	1.0000
8	1.0000
9	1.0000
10	1.0000

Figure 4.2

For each value of *switch*, the variable *lesseq* represents the cumulative probability of observing that number or fewer failures.

Example 4.1 (cont.)

We want to determine the probability that no more the 1 of the 10 switches in the sample fails inspection or, symbolically, $P(X \leq 1 | n = 10, p = 0.10)$. The variable *lesseq* tells us that the probability that no more than 1 of the 10 switches in the sample fails inspection is 0.7361, given the probability that any given switch will fail is $p = 0.10$.

Given the same values of n and p, suppose we want to find the probability that strictly less than two switches fail inspection ($P(X < 2 | n = 10, p = 0.10)$). Steps 2 – 6 can be repeated with the changes that the target variable is named *less* and the "Numeric Expression" box reads **CDF.BINOM(switch–1,10,0.10)**. If we want to determine the probability that exactly two of the 10 switches fail inspection ($P(X = 2 | n = 10, p = 0.10)$), steps 2 – 6 can be repeated with the changes that the target variable is named *equal* and the "Numeric Expression" box reads **CDF.BINOM(switch,10,0.10) – CDF.BINOM(switch–1,10,0.10)**. If we want to determine the probability that at least two of the 10 switches fail inspection ($P(X \geq 2 | n = 10, p = 0.10)$), steps 2 – 6 can be repeated with the changes that the target variable is named *greateq* and the "Numeric Expression" box reads **1 – CDF.BINOM(switch–1,10,0.10)**. Last, if we want to determine the probability that strictly more than two of the 10 switches fail inspection ($P(X > 2 | n = 10, p = 0.10)$), steps 2 – 6 can be repeated with the changes that the target variable is named *greater* and the "Numeric Expression" box reads **1 – CDF(switch,10,0.10)**.

The probabilities associated with $X \leq 2$, $X < 2$, $X = 2$, $X \geq 2$, and $X > 2$ are shown in Figure 4.3 under the variables of *lesseq*, *less*, *equal*, *greateq*, and *greater*, respectively. All five of these variables were changed so that they show four decimal places of accuracy (see Section 0.2).

switch	lesseq	less	equal	greateq	greater
2	.9298	.7361	.1937	.2639	.0702

Figure 4.3

Section 4.2. Normal Probability Distributions

Example 4.2 | The level of cholesterol in the blood is important because high cholesterol levels increase the risk of heart disease. The distribution of blood cholesterol levels in a large population of people of the same age and sex is roughly normal. For 14-year-old boys, the mean is 170 milligrams of cholesterol per deciliter of blood (mg/dl) and the standard deviation is 30 mg/dl. Using software and the normal distribution, find the approximate probability that a randomly selected 14-year-old boy has a cholesterol level less than 240 mg/dl. Find the approximate probability that a randomly selected 14-year-old boy has a cholesterol level of more than 240 mg/dl of cholesterol. Find the approximate probability that a randomly selected 14-year-old boy has a cholesterol level between 170 and 240 mg/dl.

Let X = the cholesterol level of a randomly selected 14-year-old boy. X is approximately N(170,30). To find the probabilities, follow the directions given in Example 1.10.

Example 4.2 (cont.) | The approximate probability that a randomly selected 14-year-old boy has a cholesterol level less than 240 mg/dl is 0.9902. The approximate probability that a randomly selected 14-year-old boy has a blood cholesterol level more than 240 is 0.0098. The approximate probability that a randomly selected 14-year-old boy has a cholesterol level between 170 and 240 is 0.4902.

Section 4.3. Other Probability Distributions

SPSS is capable of computing probabilities for a number of distributions. Table 4.1 displays a number of commonly used distributions and their commands.

Table 4.1

Distribution	SPSS Command
Chi-square	CDF.CHISQ(q,df)
Exponential	CDF.EXP(q,scale)
F	CDF.F(q,df1,df2)
Geometric	CDF.GEOM(q,p)
Hypergeometric	CDF.HYPER(q,total,sample,hits)
Poisson	CDF.POISSON(q,mean)
Uniform	CDF.UNIFORM(q,min,max)

Chapter 5. Sampling Distributions

A probability distribution for a sample statistic is often referred to as a **sampling distribution**. This chapter describes how to simulate random samples from a known population and compute sample statistics for the generated samples. The generated samples can then be utilized to examine properties of sampling distributions.

Section 5.1. Generating Binomial Data

Example 5.1
(IPS Ex. 5.9)

A national opinion study was conducted to determine what percent of adults agree that they like to buy new clothes, but they find shopping frustrating and time-consuming. Suppose that, in fact, 60% of the population feel this way. A study is conducted that surveys the opinion of 2500 adults. Simulation will be used to examine properties of the sampling distribution for the proportions.

To simulate 500 replications of 2500 surveys, follow these steps.
1. Click **File**, click **New**, and then click **Syntax**. The SPSS Syntax Editor in Figure 5.1 (without the text) appears.
2. Type the program appearing in Figure 5.1 into the Syntax Editor.

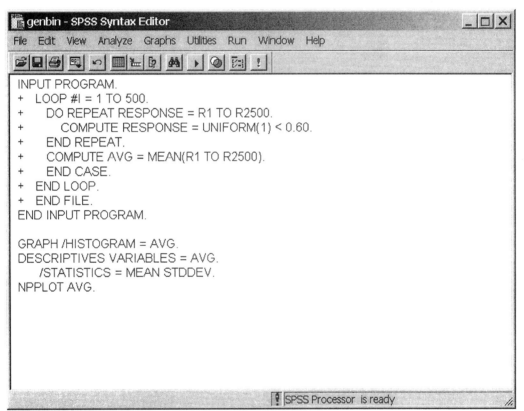

Figure 5.1

3. To run the program, click **Run**, then click **All**.

The SPSS output in Figure 5.2 and Table 5.1 displays the resulting histogram and table of descriptive statistics for a particular simulation. Note that another simulation will produce different results.

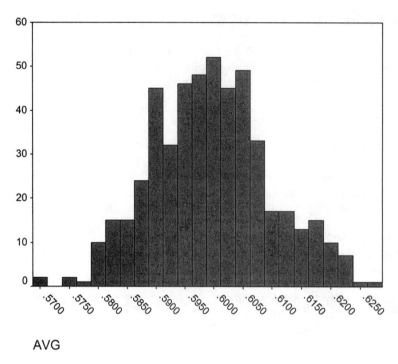

Figure 5.2

Descriptive Statistics

	N	Mean	Std. Deviation
AVG	500	.5994	.01007
Valid N (listwise)	500		

Table 5.1

Example 5.1 (cont.)

By the Central Limit Theorem, the sampling distribution for the sample proportion in this example is approximately normal with mean and standard deviation of
$$\mu_{\hat{p}} = 0.60 \text{ and } \sigma_{\hat{p}} = \sqrt{(0.60)(0.40)/2500} = 0.0098.$$
The results of the simulation are consistent with the theoretical results.

Section 5.2. Generating Normal Data

Example 5.2 (IPS Ex. 5.14)

The height of a randomly selected young woman follows a normal distribution with a mean of 64.5 inches and a standard deviation of 2.5 inches. If a medical study asked the height of 100 young women, the sampling distribution of the sample mean height would have a sampling distribution that is approximately normal with mean and standard deviation of
$$\mu_{\bar{x}} = 64.5 \text{ and } \sigma_{\bar{x}} = 2.5/\sqrt{100} = 0.25.$$

A simulation that confirms the theoretical results will be demonstrated.

To simulate 1000 replications of this study, follow these steps.
1. Click **File**, click **New**, and then click **Syntax**. The SPSS Syntax Editor in Figure 5.3 (without the text) appears.
2. Type the program appearing in Figure 5.3 into the Syntax Editor.

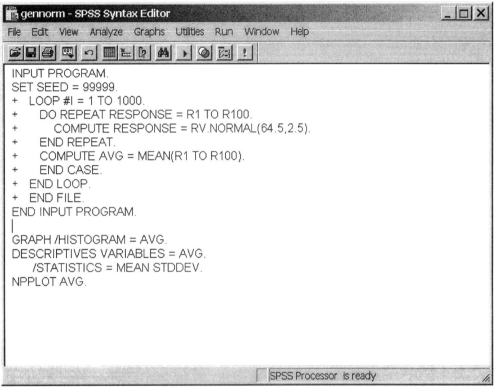

Figure 5.3

3. To run the program, click **Run**, then click **All**.

The resulting SPSS output contains the histogram, descriptive statistics, and normal Q-Q plot appearing in Figure 5.4, Table 5.2, and Figure 5.5, respectively.

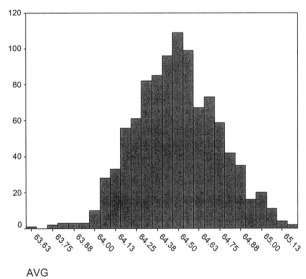

Figure 5.4

Descriptive Statistics

	N	Mean	Std. Deviation
AVG	1000	64.4943	.25016
Valid N (listwise)	1000		

Table 5.2

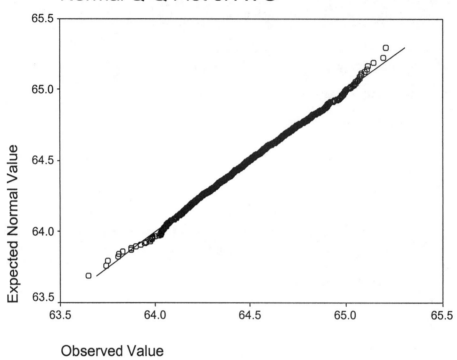

Figure 5.5

Chapter 6. Introduction to Inference

A **confidence interval** is a procedure for estimating a population parameter using observed data, and a **test of significance** is a procedure for determining the validity of a claim using observed data. This chapter describes how to **simulate** random samples from a known population and examine various properties of confidence intervals and tests of significance.

Section 6.1. Confidence Intervals

Example 6.1 (IPS Ex. 4.21)

The height of a randomly selected young woman follows a normal distribution with a mean of 64.5 inches and a standard deviation of 2.5 inches. If a medical study asked the height of 25 young women, the data might be used to estimate the mean height of all young women. Assuming that $\sigma = 2.5$, a 90% confidence interval for the mean height is given by $\bar{x} \pm 1.645\left(2.5/\sqrt{25}\right)$.

A simulation can be used to demonstrate how a confidence interval works. Specifically, the following simulation generates 1000 random samples of 25. For each sample, the mean is computed, and the resulting 90% confidence interval is constructed. The program then determines whether each confidence interval contains the true mean of 64.5. After the 1000 replications, the simulation tabulates the percentage of confidence intervals that contain the true mean.

To simulate 1000 replications of 25 samples, follow these steps.
1. Click **File**, click **New**, and then click **Syntax**. The SPSS for Windows Syntax Editor in Figure 6.1 (without the text) appears.
2. Type the program appearing in Figure 6.1 into the Syntax Editor.
3. To run the program, click **Run**, then click **All**.

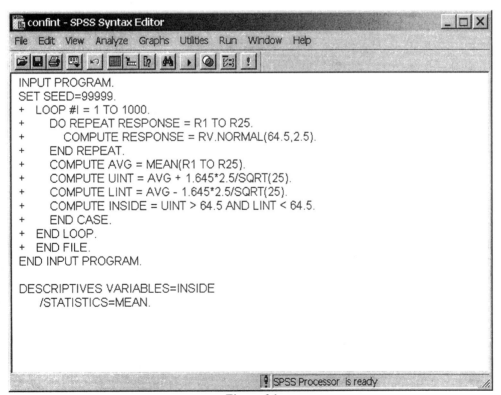

Figure 6.1

Example 6.1 (cont.) Table 6.1 displays the percentage of confidence intervals in the simulation that contain the true mean. The observed percentage of confidence intervals that contain the true mean of 64.5 is 89.3%. Note that additional simulations can be run by either setting different seeds or allowing the computer to generate a seed from the computer clock.

Descriptive Statistics

	N	Mean
INSIDE	1000	.8930
Valid N (listwise)	1000	

Table 6.1

Section 6.2. Tests of Significance

Example 6.2
(IPS Ex. 6.16
& Exer. 6.84)

Bottles of a popular cola drink are supposed to contain 300 milliliters (ml) of cola, although there is always inherent variation from bottle to bottle because the filling machinery is not precise. The distribution of the contents is normal with standard deviation $\sigma = 3$ ml. A student who suspects that the bottling company is underfilling the bottles measures the contents of six bottles and tests the hypotheses $H_0: \mu = 300$ and $H_a: \mu < 300$ at the 5% significance level.

A simulation can be used to demonstrate how a test of significance works. Specifically, the following simulation generates 1000 random samples of 6. For each sample, the mean is computed, and the resulting test statistic is determined. The program then determines whether each test statistic is less than the critical value of -1.645. After the 1000 replications the simulation tabulates the percentage of test statistics that are less than the critical value of -1.645.

To simulate 1000 replications of 6 samples, follow these steps.
1. Click **File**, click **New**, and then click **Syntax**. The SPSS for Windows Syntax Editor in Figure 6.2 (without the text) appears.
2. Type the program appearing in Figure 6.2 into the Syntax Editor.
3. To run the program, click **Run**, then click **All**.

Table 6.2 displays the percentage of test statistics that are less than the critical value of -1.645.

Descriptive Statistics

	N	Mean
INSIDE	1000	.0490
Valid N (listwise)	1000	

Table 6.2

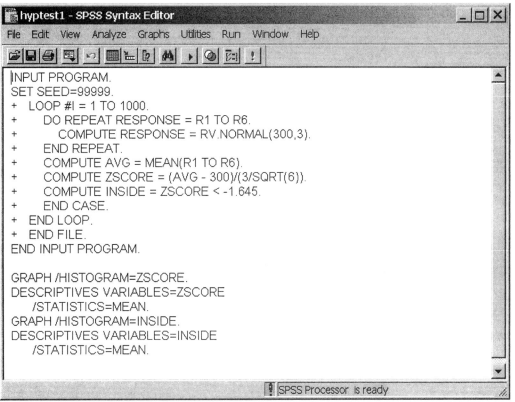

Figure 6.2

Example 6.2 (cont.)

The observed significance level for the simulation was 0.049, which is consistent with the significance level of 0.05. In addition, a simulation can be used to examine the power of the test against the specific alternative of $\mu = 297$. The test rejects H_o when the test statistic z is less than -1.645 or when the sample mean $\bar{x}$ is less than 297.985. Then the power is

$$P(\bar{x} \leq 297.985) = P\left(z \leq \frac{297.985 - 297}{3/\sqrt{6}}\right) = P(z \leq 0.8045) = 0.7894.$$

The previous program can be used to simulate the observed power by generating samples with a mean of 297. This is accomplished by generating a RESPONSE = RV.NORMAL(297,3) rather than a RESPONSE = RV.NORMAL(300,3). Table 6.3 displays the results of the revised simulation. The observed power for the simulation was 0.7790, which is consistent with the power of the test against the specific alternative of $\mu = 297$.

Descriptive Statistics

	N	Mean
INSIDE	1000	.7790
Valid N (listwise)	1000	

Table 6.3

Chapter 7. Inference for Distributions

Section 7.1. Inference for the Mean of a Population

This section introduces the use of the *t* **distribution** in inferential statistics for the mean of a population. When σ is unknown for the population, the *t* distribution, rather than the *z* distribution, is used. The particular *t* distribution is specified by giving the degrees of freedom. The **one-sample *t* confidence interval for** μ (the population mean), the **one-sample *t* test for** μ, and the **matched pairs *t* procedures** are discussed in this section.

<u>One-Sample *t* Confidence Interval</u>

Example 7.1
(IPS Ex. 7.1)

In fiscal year 1996, the U.S. Agency for International Development provided 238,300 metric tons of corn soy blend (CSB) for development programs and emergency relief in countries throughout the world. CSB is highly nutritious, low-cost fortified food that is partially pre-cooked and can be incorporated into different food preparations by the recipients. As part of a study to evaluate appropriate vitamin C levels in this commodity, measurements were taken on samples of CSB produced in a factory. The following data are the amounts of vitamin C, measured in milligrams per 100 grams of blend, for a random sample of size 8 from a production run: 26, 31, 23, 22, 11, 22, 14, 31.

Compute a 95% confidence interval for μ where μ is the mean vitamin C content of the CSB. Before proceeding with the confidence interval for the mean, we must verify the assumption of normally distributed data. We cannot effectively check normality with only 8 observations, but a stemplot can be used to verify that data set has no outliers. Using the SPSS for Windows Data Editor, one can enter a single variable called *vitaminc*, declared as numeric 8.2.

To obtain a confidence interval for μ, follow these steps.
1. Click **Analyze**, click **Descriptive Statistics**, and then click **Explore**. The "Explore" window in Figure 7.1 appears.
2. Click *vitaminc*, then click ▶ to move *vitaminc* into the "Dependent List" box.
3. By default, a 95% confidence interval for μ will be computed. To change the confidence level, click **Statistics**. Change 95 in the "Confidence Interval for Mean" box to the desired confidence level. Click **Continue**.
4. By default, the "Display" box in the lower left corner of the "Explore" window has "Both" selected. Click **Statistics**. If you want to obtain a stemplot to check the assumption of normality, then skip this step.
5. Click **OK**.

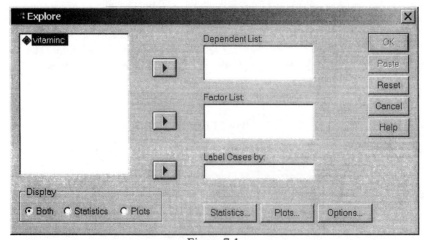

Figure 7.1

Table 7.1 contains the resulting SPSS for Windows output.

Descriptives

			Statistic	Std. Error
VITAMINC	Mean		22.5000	2.54250
	95% Confidence Interval for Mean	Lower Bound	16.4880	
		Upper Bound	28.5120	
	5% Trimmed Mean		22.6667	
	Median		22.5000	
	Variance		51.714	
	Std. Deviation		7.19126	
	Minimum		11.00	
	Maximum		31.00	
	Range		20.00	
	Interquartile Range		13.7500	
	Skewness		-.443	.752
	Kurtosis		-.631	1.481

Table 7.1

Example 7.1 (cont.) We are 95% confident that the mean vitamin C content of the CSB for this run is between 16.488 and 28.512 mg/100g (obtained from the "Lower Bound" and "Upper Bound" rows in Table 7.1).

One-Sample *t* Test

Example 7.2 (IPS Ex. 7.3) Suppose that we know that sufficient vitamin C was added to the CSB mixture to produce a mean vitamin C content in the final product of 40 mg/100 g. It is suspected that some of the vitamin is lost or destroyed in the production process. To test this hypothesis we can conduct a one-sided test to determine if there is sufficient evidence to conclude that the CSB mixture lost vitamin C content at $\alpha = 0.05$ level.

Before proceeding with the one-sample *t* test, we must verify the assumption of normally distributed data. We cannot effectively check the normality assumption with only 8 observations, but a stemplot can be used to verify that this data has no gaps or outliers or other signs of nonnormal behavior. Set up $H_0: \mu = 40$ versus $H_a: \mu < 40$, where μ = the mean vitamin C content of the final product. From the previous example, the SPSS for Windows Data Editor contains a single variable called ***vitaminc***, which is declared type numeric 8.2.

To conduct a one-sample *t* test, follow these steps.
1. Click **Analyze**, click **Compare Means**, and then click **One-Sample T Test**. The "One-Sample T Test" window in Figure 7.2 appears.
2. Click ***vitaminc***, then click ▸ to move ***vitaminc*** to the "Test Variable(s)" box.
3. Type the value of μ_0 (the value of μ under H_0) into the "Test Value" box. For this example, one must type "40" in the "Test Value" box.
4. By default, a 95% confidence interval for $\mu - \mu_0$ will be part of the one-sample *t* test output. To change the confidence level, click **Options**, change 95 in the "Confidence Interval" box to the desired confidence level, and then click **Continue**.
5. Click **OK**.

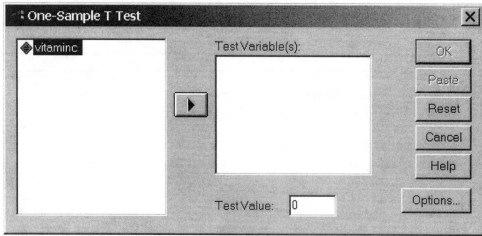

Figure 7.2

Table 7.2 is part of the resulting SPSS for Windows output.

One-Sample Test

	Test Value = 40					
					95% Confidence Interval of the Difference	
	t	df	Sig. (2-tailed)	Mean Difference	Lower	Upper
VITAMINC	-6.883	7	.000	-17.5000	-23.5120	-11.4880

Table 7.2

Example 7.2 (cont.)

Because of the one-sided alternative, we are only interested in the area below the test statistic $t = -6.883$. This P-value is 0.000. Ordinarily, the one-sided P-value would be obtained by taking the value under the "Sig. (2-tailed)" column and dividing it by 2 (0.000/2). We can safely reject H_0 in favor of H_a; that is, there is sufficient evidence at the 0.05 level to conclude that the mean vitamin C content is less than 40 mg/100 g. In other words, there is strong evidence to conclude that some of the vitamin C content has been lost or destroyed. In addition, we are 95% confident that the true mean sweetness loss lies between 16.488 and 28.512. Note that when the test value $\mu_0 \neq 0$, then the confidence interval for μ can be obtained by adding μ_0 to the values located in the "Lower" and "Upper" boxes within the "Confidence Interval of the Difference" box (see Table 7.2).

The next example shows how to generate the exact P-value for the one-sample t test when summarized data rather than raw data have been provided.

Example 7.3

The one-sample t statistic for testing H_0: $\mu = 10$ versus one of the following, H_a: $\mu < 10$, H_a: $\mu > 10$, or H_a: $\mu \neq 10$ from a sample of $n = 23$ observations has a test statistic value of $t = 2.78$. Using software, find the exact P-value.

To obtain the P-value, follow these steps.
1. Enter the value of the test statistic into the SPSS for Windows Data Editor under the variable called ***teststat***.
2. From the SPSS for Windows main menu bar, click **Transform** and then click **Compute**. The "Compute Variable" window in Figure 7.3 appears.

3. In the "Target Variable" box, type *pvalue*.
4. In the "Functions" box, click ▼ until **CDF.T(q,df)** appears in the "Functions" box. Double click on **CDF.T(q,df)** to move **CDF.T(?,?)** into the "Numeric Expression" box. The CDF.T(q,df) function stands for the cumulative distribution function for the *t* distribution, and it calculates the area to the left of *q* under the correct *t* distribution.

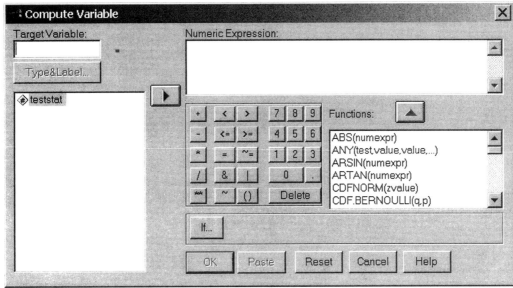

Figure 7.3

5. In the "Numeric Expression" box, change the first **?** to *teststat* and the second **?** to **22** (23 − 1). For H_a: $\mu < 10$, **CDF.T(*teststat*,22)** should appear in the "Numeric Expression" box. For H_a: $\mu > 10$, **1 − CDF.T(*teststat*,22)** should appear in the "Numeric Expression" box. For H_a: $\mu \neq 10$, **2*(1 − CDF.T(ABS(*teststat*),22))** should appear in the "Numeric Expression" box.
6. Click **OK**.

Example 7.3 (cont.) The *P*-value can be found in the SPSS for Windows Data Editor. By default, the number of decimal places for the variable *pvalue* is two. The number of digits after the decimal place can be changed by following the directions in Section 0.2. The *P*-value for H_a: $\mu < 10$ is 0.995. The *P*-value for H_a: $\mu > 10$ is 0.005. The *P*-value for H_a: $\mu \neq 10$ is 0.011.

Matched Pairs *t* Procedure

Example 7.4 (IPS Ex. 7.7) The National Endowment for the Humanities sponsors summer institutes to improve the skills of high school teachers for foreign languages. One such institute hosted 20 French teachers for 4 weeks. At the beginning of the period, the teachers were given the Modern Language Association's listening test of understanding spoken French. After 4 weeks of immersion in French in and out of class, the listening test was given again. Table 7.3 gives the pretest and posttest scores. The maximum possible score is 36. Assess whether the institute significantly improved the teachers' comprehension of spoken French at the $\alpha = 0.10$.

This example is a **matched pairs study**, in which repeated measurements on the same subjects were obtained. To obtain the SPSS for Windows output presented in this example (especially the matched pairs *t* test output), the pretest score (*pretest*) was the first variable in the SPSS for Windows Data Editor and the posttest score (*posttest*) was the second variable, where both variables were declared numeric 8.2.

Table 7.3 Modern Language Association listening scores for French teachers

Subj.	Pretest	Posttest	Gain	Subj.	Pretest	Posttest	Gain
1	32	34	2	11	30	36	6
2	31	31	0	12	20	26	6
3	29	35	6	13	24	27	3
4	10	16	6	14	24	24	0
5	30	33	3	15	31	32	1
6	33	36	3	16	30	31	1
7	22	24	2	17	15	15	0
8	25	28	3	18	32	34	2
9	32	26	–6	19	23	26	3
10	20	26	6	20	23	26	3

Example 7.4 (cont.) Before proceeding with the matched pairs *t* test, we must verify the assumption that the differences (the Gain column in Table 7.3, which equals *posttest – pretest*) come from a normal distribution. The first teacher (or subject), for example, improved from 32 to 34, so the gain is 34 – 32 = 2. Because higher scores represent better performance, positive differences show that the subject did better after attending the institute. If the differences have not been entered into the original SPSS for Windows data set, then a variable *gain* must be created to check the assumption of normality.

To create the variable *gain*, follow these steps.
1. Click **Transform** and then click **Compute**. The "Compute Variable" window in Figure 7.3 appears, with the exception that *pretest* and *posttest* appear in the window.
2. In the "Target Variable" box, type *gain*.
3. Double click *posttest*, click the gray minus sign (–), and then double click *pretest*. The expression *posttest – pretest* appears in the "Numeric Expression" box.
4. Click **OK**. The variable *gain* appears in the SPSS for Windows Data Editor.

Example 7.4 (cont.) To obtain the normal quantile plot for the variable *gain*, see the directions given in Example 1.11. Even though the normal quantile plot displays an outlier at – 6, the overall pattern of the plot is otherwise roughly straight. For this reason, the matched pairs *t* test will be applied to this data. Note that the textbook addresses this concern in its section on the robustness of the *t* test.

Set up H_0: $\mu = 0$ versus H_a: $\mu > 0$, where μ = the mean gain in the population of French teachers if all of them attended the institute. The null hypothesis says that no improvement occurs, and the alternative hypothesis says that posttest scores, on average, exceed the pretest scores.

To conduct a matched pairs *t* test, follow these steps.
1. Click **Analyze**, click **Compare Means**, and then click **Paired–Samples T Test**. The "Paired–Samples T Test" window in Figure 7.4 appears.
2. Click *posttest*. The variable *posttest* appears after "Variable 1" in the "Current Selections" box.
3. Click *pretest*. The variable *pretest* appears after "Variable 2" in the "Current Selections" box.
4. Click ▶. The expression *posttest – pretest* appears in the "Paired Variables" box.
5. By default, a 95% confidence interval for μ will be part of the matched pairs *t* test output. To change the confidence level, click **Options**, change 95 to **90** in the "Confidence Interval" box, and then click **Continue**.
6. Click **OK**.

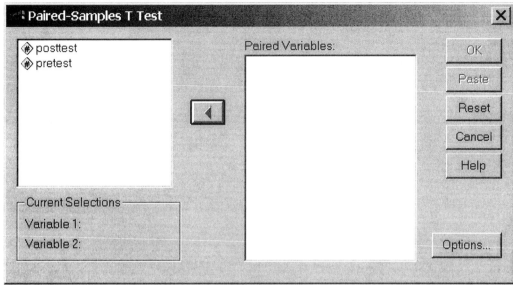

Figure 7.4

Table 7.4 is part of the resulting SPSS for Windows output.

Paired Samples Test

	Paired Differences							
				90% Confidence Interval of the Difference				
	Mean	Std. Deviation	Std. Error Mean	Lower	Upper	t	df	Sig. (2-tailed)
Pair 1 POSTTEST - PRETEST	2.5000	2.89282	.64685	1.3815	3.6185	3.865	19	.001

Table 7.4

NOTE: The same results would have been obtained if we had applied the one-sample t test to the variable ***gain*** using a test value of 0 (follow the directions given in Example 7.2).

Example 7.4 (cont.)

Because of the one-sided alternative, we are only interested in the upper right tail above $t = 3.865$. Therefore, the P-value is 0.0005. This is obtained by taking the value under the "Sig. (2-tailed)" column and dividing it by 2 (0.001/2). The data do support the claim that the institutes improve mean scores. This improvement is statistically significant at the 10% level. In addition, we are 90% confident that the mean improvement that would be achieved if the entire population of French teachers attended a summer institute lies somewhere between 1.3815 and 3.6185. Though statistically significant, the effect of the institute was rather small.

If we drop the single outlier of -6, the P-value would change from 0.0005 to 0.0000058. The results of the t procedure with the outlier are conservative in the sense that the conclusions show a smaller effect than would be the case if the outlier was not present.

Inference for Nonnormal Populations Using the Sign Test

Example 7.5
(IPS Ex. 7.12)

Return to the data of Example 7.4 showing the improvement in French listening scores after attending a summer institute. In that example we used the matched pairs t test on these data, despite an outlier that makes the P-value only roughly correct. The sign test is based on the following simple observation: of the 17 teachers whose scores changed, 16 improved and only 1 did worse. The sign test is a distribution-free procedure designed to cope with nonnormal data. The distribution-free test does not ask the same question (Has the mean changed?) that the t test asks. Set up $H_0: p = \frac{1}{2}$ versus $H_a: p > \frac{1}{2}$, where p = the probability that a randomly chosen teacher would improve if she attended the institute. The P-value calculation is based on the binomial distribution rather than the t distribution.

To obtain the SPSS output presented in this example, the posttest score was the first variable in the SPSS data set and the pretest score was the second variable, where both variables were declared numeric 8.2.

To perform the sign test, follow these steps.
1. Click **Analyze**, click **Nonparametric Tests**, and then click **2 Related Samples**. The SPSS window in Figure 7.5 appears.
2. Click *posttest*. The variable *posttest* appears after "Variable 1" in the "Current Selections" box.
3. Click *pretest*. The variable *pretest* appears after "Variable 2" in the "Current Selections" box.
4. Click ▶. Then *posttest – pretest* appears in the "Test Pair(s) List" box.
5. Click **Wilcoxon** in the "Test Type" box. The ✓ in front of "Wilcoxon" disappears.
6. Click **Sign** in the "Test Type" box. A ✓ appears in front of "Sign".
7. Click **OK**.

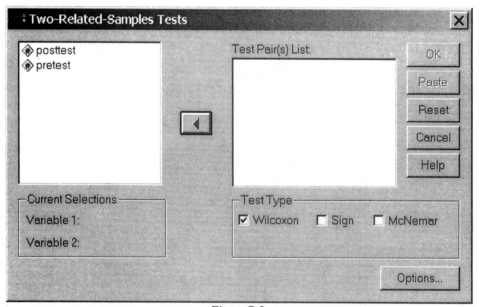

Figure 7.5

Tables 7.5 and 7.6 are the resulting SPSS output. Although *posttest – pretest* had appeared in the "Test Pair(s) List" box, SPSS gives output involving *pretest – posttest*. The output states that the negative and positive differences are 16 and 1, respectively, for *pretest – posttest*. Thus, the negative and positive differences would be 1 and 16, respectively, for *posttest – pretest*. However, the two-tailed P-value (located in the Exact Sig. (2-tailed) value box) is the same whether *pretest – posttest* or *posttest – pretest* is used.

Frequencies

		N
PRETEST - POSTTEST	Negative Differences[a]	16
	Positive Differences[b]	1
	Ties[c]	3
	Total	20

a. PRETEST < POSTTEST
b. PRETEST > POSTTEST
c. POSTTEST = PRETEST

Table 7.5

Test Statistics[b]

	PRETEST - POSTTEST
Exact Sig. (2-tailed)	.000[a]

a. Binomial distribution used.
b. Sign Test

Table 7.6

Example 7.5 (cont.) The Exact Sig. (2-tailed) value box contains the P-value for H_a: $p \neq \frac{1}{2}$. The value of 0.000 does not mean that the two-tailed P-value is zero; it means that the two-tailed P-value is less than 0.001. Because of the one-sided alternative, we can safely say that the P-value is less than 0.0005 (0.001/2). Therefore, we can reject H_0 in favor of H_a; that is, there is sufficient evidence at the 0.10 level to conclude that the probability that a randomly chosen teacher would improve if she attended the institute is greater than 1/2.

Section 7.2. Two-Sample *t* Procedures

This section introduces the use of the *t* distribution in inferential statistics for comparing two means. The two-sample problem examined in this section compares the responses in the two groups, where the responses in each group are independent of those in the other group. Assuming the two samples come from normal populations, the **two-sample *t* procedure** is the correct test to apply.

Example 7.6 (IPS Ex. 7.17) The pesticide DDT causes tremors and convulsions if it is ingested by humans or other mammals. Researchers seek to understand how the convulsions are caused. In a randomized comparative experiment, 6 white rats poisoned with DDT were compared with a control group of 6 unpoisoned rats. Electrical measurements of nerve activity are the main clue to the nature of DDT poisoning. When a nerve is stimulated, its electrical response shows a sharp spike followed by a much smaller second spike. Researchers found that the second spike is larger in rats fed DDT than in normal rats. This observation helps biologists understand how DDT causes tremors. The researchers measured the amplitude of the second spike as a percentage of the first spike when a nerve in the rat's leg was stimulated. Is there sufficient evidence at the 0.05 level to conclude that the population means differ between the two groups? The data for the 12 rats are given in Table 7.7.

Table 7.7 DDT versus Control

Rat	Group	Spike	Rat	Group	Spike
1	DDT	12.207	7	CONTROL	11.074
2	DDT	16.869	8	CONTROL	9.686
3	DDT	25.050	9	CONTROL	12.064
4	DDT	22.429	10	CONTROL	9.351
5	DDT	8.456	11	CONTROL	8.182
6	DDT	20.589	12	CONTROL	6.642

Example 7.6 (cont.) Before proceeding with the two-sample t test, we must verify that the assumptions of normally distributed data in both groups are reasonably satisfied. Both populations are reasonably normal (which can be determined using normal quantile plots for the two groups of data; see directions in Example 1.11), as far as can be judged from six observations.

Set up H_0: $\mu_1 = \mu_2$ versus H_a: $\mu_1 \neq \mu_2$, where μ_1 = the true mean percentage of the amplitude of the second spike compared to the first spike in the DDT group and μ_2 = the true mean percentage of the amplitude of the second spike compared to the first spike in the control group. The SPSS Data Editor contains the variables *group* (declared type string 8) and *spike* (declared numeric 8.3).

To perform the two-sample t test, follow these steps.
1. Click **Analyze**, click **Compare Means**, and then click **Independent-Samples T Test**. The SPSS window in Figure 7.6 appears.

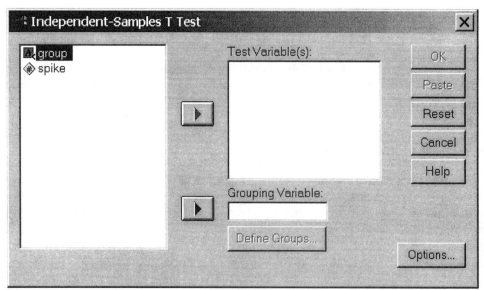

Figure 7.6

2. Click *spike*, then click ▶ to move *spike* into the "Test Variable(s)" box.
3. Click *group*, then click ▶ to move *group* into the "Grouping Variable" box.
4. Click **Define Groups**, and an SPSS window appears.
5. Type ***DDT*** in the "Group 1" box. Press the tab key. Type ***CONTROL*** in the "Group 2" box. After you have typed in the group names (which must typed exactly as they appear in the SPSS Data Editor). The SPSS window will appear as displayed in Figure 7.7. Click **Continue**. (NOTE: **Group('DDT' 'CONTROL')** will appear in the "Grouping Variable" box.)

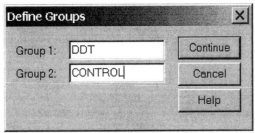

Figure 7.7

6. By default, a 95% confidence interval for $\mu_1 - \mu_2$ (the difference in population means) will be part of the two-sample t test output. To change the confidence level, click **Options**, change 95 to the desired confidence level in the "Confidence Interval" box, and then click **Continue**.
7. Click **OK**.

Table 7.8 is part of the resulting SPSS output.

Independent Samples Test

		Levene's Test for Equality of Variances		t-test for Equality of Means					95% Confidence Interval of the Difference	
		F	Sig.	t	df	Sig. (2-tailed)	Mean Difference	Std. Error Difference	Lower	Upper
SPIKE	Equal variances assumed	7.658	.020	2.991	10	.014	8.1002	2.70802	2.06632	14.13401
	Equal variances not assumed			2.991	5.938	.025	8.1002	2.70802	1.45697	14.74336

Table 7.8

Example 7.6 (cont.)

SPSS reports the results of two t procedures: the pooled two-sample t procedure (assumes equal population variances) and a general two-sample t procedure (does not assume equal population variances). To determine which t procedure to use, SPSS performs Levene's Test for Equality of Variances for H_0: $\sigma_1^2 = \sigma_2^2$ versus H_a: $\sigma_1^2 \neq \sigma_2^2$. The F test statistic for the Levene's test is obtained by computing a one-way analysis of variance (see Chapter 12) on the absolute deviations of each case from its group mean. The P-value for Levene's test of 0.020 is located under the Sig. column. We reject H_0 in favor of H_a. That is, there is sufficient evidence at the 0.05 level to conclude that the population variances are unequal. Thus, we use the two-sample t procedure for which equal variances are not assumed. The P-value for the appropriate two-sample t procedure is 0.025 (found under the Sig. (2-tailed) column). We reject H_0 in favor of H_a; that is, there is sufficient evidence at the 0.05 level to conclude that the population means differ between the two groups. In addition, we are 95% confident that $\mu_1 - \mu_2$ lies somewhere between 1.457 and 14.743.

Chapter 8. Inference for Proportions

This chapter in IPS focuses on inference about population proportions. The first section in this chapter addresses inference on a single population proportion, and the second section addresses comparing two population proportions. Although SPSS programs could be written to perform inference for proportions, it is the our opinion that the analyses discussed in this chapter can be better accomplished using a calculator or another statistical package. It should be noted that after the test statistic is computed for a test of significance, SPSS can be used to easily compute the *P*-value for the test as described in Example 7.3 of Section 7.1. Note that the problems in this chapter utilize the z distribution. As a result, the *P*-values would be computed using the function CDF.NORMAL(q,0,1), as described in Example 1.10 of Section 1.3.

Chapter 9. Inference for Two-Way Tables

This chapter introduces the notion of describing the relationship between two categorical variables. To analyze categorical data, we use counts (frequencies) or percents (relative frequencies) of observations that fall into various categories. The data are often summarized using a **two-way table**, with the two referring to the number of variables. **Inferential** procedures are applied to test whether the two categorical variables are independent.

Example 9.1
(IPS Ex. 9.12)

Do men and women participate in sports for the same reasons? A study on why students participate in sports collected data from 67 male and 67 female undergraduates at a large university. Each student was classified into one of four categories based on his or her responses to a questionnaire about sports goals. The four categories were high social comparison-high mastery (HSC-HM), high social comparison-low mastery (HSC-LM), low social comparison-high mastery (LSC-HM), and low social comparison-low mastery (LSC-LM). One purpose of the study was to compare the goals of male and female students. The data are displayed in a two-way table (Table 9.1). Since there are 4 levels of the row variable (***goal***) and there are two levels of the column variable (***gender***), this is referred to as a 2 × 4 contingency table. As such, it has eight cells, where each cell represents a unique combination of a row level and a column level.

The entries in this table are the observed, or sample, counts. For example, there are 14 females in the high social comparison-high mastery group. Note that the marginal totals are also given in the table. They are not part of the raw data but are calculated by summing the rows or columns. The column totals are the numbers of observations sampled in both populations. The grand total, 134, can be obtained by summing the row or column totals. It is the total number of observations in the study.

Table 9.1 Counts of Goal by Gender

	Observed Counts for Sports Goal		
	Gender		
Goal	Female	Male	Total
HSC-HM	14	31	45
HSC-LM	7	18	25
LSC-HM	21	5	26
LSC-LM	25	13	38
Total	67	67	134

Since these data were provided as summary counts (frequencies), there is a short-cut way of entering these data into SPSS. This method employs the use of a third variable that will be used as a weighting variable. The variables of ***gender***, ***goal***, and ***weight*** were all declared numeric variables of length 8.0. The values of "1" and "2" were used to represent the genders of "Female" and "Male", respectively. Further, the values of "1", "2", "3", and "4" were used to represent the goals of "HSC-HM", "HSC-LM", "LSC-HM", and "LSC-LM", respectively. Value labels were assigned for the variables of ***gender*** and ***goal***. Note that the values associated with the weight variable are merely the counts from each of the corresponding eight cells. Using the weight variable significantly reduces data entry for this problem (only eight rows are needed). If the weight variable were not used, it would take 134 rows to enter the same information (14 rows which contained the values of "Female" and "HSC-HM" for the variables of ***gender*** and ***goal***, respectively; 7 rows which contained the values of "Female" and "HSC-LM"; etc).

Figure 9.1 shows how the data were entered into SPSS.

gender	goal	weight
1	1	14
1	2	7
1	3	21
1	4	25
2	1	31
2	2	18
2	3	5
2	4	13

Figure 9.1

To take advantage of this data-entry shortcut, however, you must first activate the weighting option. To activate the weighting option, follow these steps.
1. Click **Data**, and click **Weight Cases**. The "Weight Cases" window in Figure 9.2 appears.

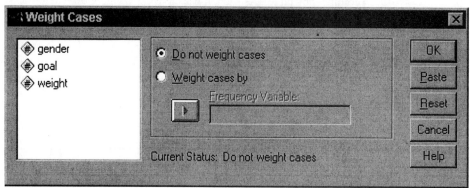

Figure 9.2

2. Click **Weight cases by**.
3. Click *weight*, then click ▸ to move *weight* into the "Frequency Variable:" box.
4. Click **OK**.

Now that the weighting option has been activated, you can generate a table similar to Table 9.1 by following these steps.
1. Click **Analyze**, click **Descriptive Statistics**, and then click **Crosstabs**. The "Crosstabs" window in Figure 9.3 appears.
2. Click *goal*, then click ▸ to move *goal* into the "Row(s)" box.
3. Click *gender*, then click ▸ to move *gender* into the "Column(s)" box.
4. Click **OK**.

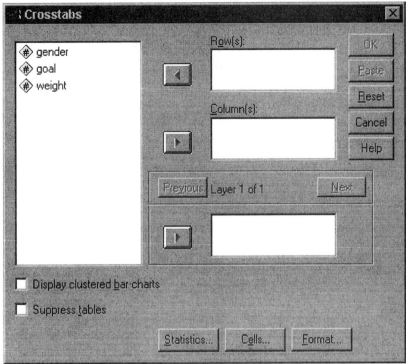

Figure 9.3

The resulting SPSS output is shown in Table 9.2.

GOAL * GENDER Crosstabulation

Count

		GENDER		
		Female	Male	Total
GOAL	HSC-HM	14	31	45
	HSC-LM	7	18	25
	LSC-HM	21	5	26
	LSC-LM	25	13	38
Total		67	67	134

Table 9.2

To describe the relationship between these two categorical variables, it is useful to generate percents. Each cell can be expressed as a percent of the grand total, the total of the row in which it is found, and the total of the column in which it is found. To generate these percents, follow these steps. **NOTE:** If you generated Table 9.2 in this session, steps 2 and 3 have already been completed.

1. Click **Analyze**, click **Descriptive Statistics**, and then click **Crosstabs**. The "Crosstabs" window in Figure 9.3 appears.
2. Click *goal*, then click ▶ to move *goal* into the "Row(s)" box.
3. Click *gender*, then click ▶ to move *gender* into the "Column(s)" box.
4. Click the **Cells** button. The "Crosstabs: Cell Display" window in Figure 9.4 appears.

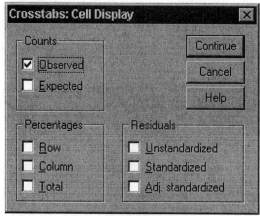

Figure 9.4

5. Click **Row**, **Column**, and **Total** within the "Percentages" area so that a check mark (✓) appears before each type of percentage.
6. Click **Continue**.
7. Click **OK**.

The resulting SPSS output is shown in Table 9.3.

GOAL * GENDER Crosstabulation

			GENDER		Total
			Female	Male	
GOAL	HSC-HM	Count	14	31	45
		% within GOAL	31.1%	68.9%	100.0%
		% within GENDER	20.9%	46.3%	33.6%
		% of Total	10.4%	23.1%	33.6%
	HSC-LM	Count	7	18	25
		% within GOAL	28.0%	72.0%	100.0%
		% within GENDER	10.4%	26.9%	18.7%
		% of Total	5.2%	13.4%	18.7%
	LSC-HM	Count	21	5	26
		% within GOAL	80.8%	19.2%	100.0%
		% within GENDER	31.3%	7.5%	19.4%
		% of Total	15.7%	3.7%	19.4%
	LSC-LM	Count	25	13	38
		% within GOAL	65.8%	34.2%	100.0%
		% within GENDER	37.3%	19.4%	28.4%
		% of Total	18.7%	9.7%	28.4%
Total		Count	67	67	134
		% within GOAL	50.0%	50.0%	100.0%
		% within GENDER	100.0%	100.0%	100.0%
		% of Total	50.0%	50.0%	100.0%

Table 9.3

Example 9.1 (cont.)	Each cell contains four entries, which are labeled at the beginning of each row. The "Count" is the cell count. The "% within GOAL" is the cell count expressed as a percent relative to the row total. As an example, out of all 45 individuals who were classified as HSC-HM, 14, or 31.1%, were female. The "% within GENDER" is the cell count expressed as a percent relative to the column total. For instance, out of the 67 males, 5, or 7.5%, were classified as LSC-HM. The "% of Total" is the cell count expressed as a percent relative to the grand total. Of the 134 total individuals who participated in the study, 25, or 18.7%, were females classified as LSC-LM.
	In this example, we are most interested in the effect of gender on the distribution of sports goals (or the conditional distributions of goals within genders). As such, to compare the genders, the column percents or the percents within gender are examined. A higher percent of males were classified as HSC, regardless of the level of mastery, whereas a higher percent of females were classified as LSC, regardless of the level of mastery. This suggests that females and males have different goals when they participate in recreational sports.
	The differences between the distribution of males and females sports goals in the sample appear to be large. A Chi-square test can be used to assess the extent to which these differences can be plausibly attributed to chance. The null and alternative hypotheses are H_0: there is no association between gender and sports goals, and H_a: there is an association between gender and sports goals.

A χ^2 test of independence can be performed to see if the relationship between two categorical variables is significant. To perform the χ^2 test of independence, follow these steps. **NOTE:** If you have generated either Table 9.2 or Table 9.3 in this session, steps 3 and 4 have already been completed.

1. Go to the Data View tab in the Data Editor.
2. Click **Analyze**, click **Descriptive Statistics**, and then click **Crosstabs**.
3. Click *goal*, then click ▶ to move *goal* into the "Row(s)" box.
4. Click *gender*, then click ▶ to move *gender* into the "Column(s)" box.
5. If you are interested in including the expected cell counts in the contingency table, click the **Cells** button. The "Crosstabs: Cell Display" window appears (see Figure 9.4). Click **Expected** under **Counts** so that a check mark (✔) appears before **Expected**. If check marks appear before **Row**, **Column**, and **Total** under the "Percentages" box, click on each of these terms so that the check marks disappear. Then click **Continue**.
6. Click the **Statistics** button.
7. Click **Chi-Square**.
8. Click **Continue**.
9. Click **OK**.

The resulting SPSS output is shown in Tables 9.4 and 9.5.

Chi-Square Tests

	Value	df	Asymp. Sig. (2-sided)
Pearson Chi-Square	24.898[a]	3	.000
Likelihood Ratio	26.036	3	.000
Linear-by-Linear Association	16.225	1	.000
N of Valid Cases	134		

a. 0 cells (.0%) have expected count less than 5. The minimum expected count is 12.50.

Table 9.4

GOAL * GENDER Crosstabulation

			GENDER		Total
			Female	Male	
GOAL	HSC-HM	Count	14	31	45
		Expected Count	22.5	22.5	45.0
	HSC-LM	Count	7	18	25
		Expected Count	12.5	12.5	25.0
	LSC-HM	Count	21	5	26
		Expected Count	13.0	13.0	26.0
	LSC-LM	Count	25	13	38
		Expected Count	19.0	19.0	38.0
Total		Count	67	67	134
		Expected Count	67.0	67.0	134.0

Table 9.5

Example 9.1 (cont.) All of the expected cell counts are moderately large, so the χ^2 distribution should reasonably approximate the *P*-value. The test statistic value (Pearson Chi-square) is $\chi^{2*} = 24.898$ with $df = 3$ and a *P*-value (Asymptotic Significance) less than 0.001. The Chi-square test confirms that the data contain clear evidence against the null hypothesis that females and male students have the same distribution of sports goals. Under H_0, the chance of obtaining a value of χ^2 greater than or equal to the calculated value of 24.898 is very small — less than 0.001.

As was mentioned previously, we are most interested in the effect of gender on the distribution of sports goals, or the conditional distributions of goals within genders. Bar graphs can help make the conditional distribution more apparent. If you are interested in obtaining a bar graph showing the conditional distribution of goals given that a person is female, follow these steps.
1. Go to the Data View tab in the Data Editor.
2. Click **Data**, and click **Select Cases**. The "Select Cases" window in Figure 9.5 appears.

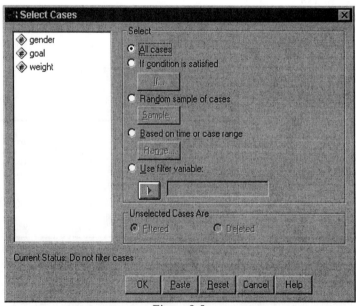

Figure 9.5

3. Click **If condition is satisfied**, then click the **If** button.
4. Click *gender*, then click ▶ to move *gender* into the upper right box.
5. Click the "=" button and then click the "**1**" button. The upper right box should now read as shown in Figure 9.6.
6. Click **Continue**, then click **OK**.

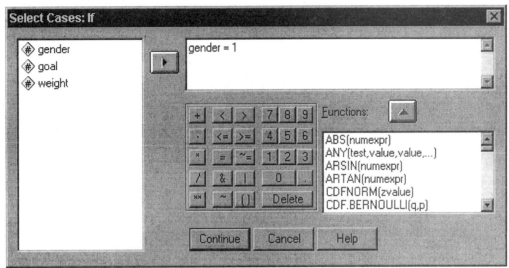

Figure 9.6

7. A new variable, *filter_$*, appears in the SPSS Data Editor. All cases that have a value of 1 for *gender* also have a value of 1 for *filter_$* (indicating that these cases were selected). In addition, all cases that have a value of 2 for *gender* have a value of 0 for *filter_$* (indicating that these cases were not selected). Also note that the unselected cases have an off-diagonal line in the case number cell.
8. Click **Graphs**, then click **Bar**. The "Bar Charts" window appears (see Figure 9.7).

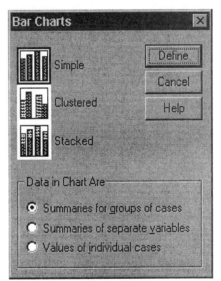

Figure 9.7

9. By default, SPSS assumes that you want a simple bar chart and that the data in the charts represent summaries for groups of cases. Since both of these are appropriate for this example, click **Define**. The "Define Simple Bar: Summaries for Group of Cases" window appears (see Figure 9.8).

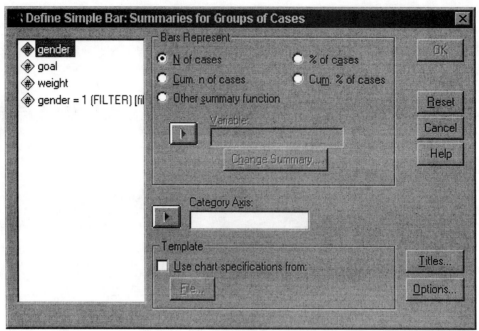

Figure 9.8

10. Click *goal*, then click ▶ to move *goal* into the "Category Axis:" box.
11. Click **% of cases** under the "Bars Represent" box.
12. Click **Titles** button. The "Titles" window appears. Type **Conditional distribution of goal** in the "Line 1" box under Title, press the **Tab** key, and type **for females** in the "Line 2" box under Title. The window should now look like Figure 9.9.

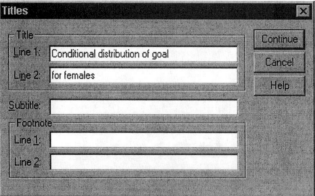

Figure 9.9

13. Click **Continue**, then click **OK**. The bar chart shown in Figure 9.10 appears (after changing the color of the bar chart, adding percent values to the bars, and centering the title).

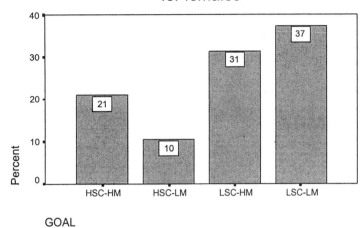

Figure 9.10

Example 9.1 (cont.)

Figure 9.10 suggests that for females, approximately 21% had a goal of high social comparison-high mastery (HSC-HM), 10% had a goal of high social comparison-low mastery (HSC-LM), 31% had a goal of low social comparison-high mastery (LSC-HM), and 37% had a goal of low social comparison-low mastery (LSC-LM). These numbers agree with the "% within gender" values (with rounding) for the variable *goal* under column one (female) in Table 9.3.

A similar bar graph can be generated for males by repeating the prior set of steps with the following changes: (1) in the "Select Cases: If" window, change the text box so it reads "gender = 2" and, (2) change the title of the graph so it reads "Conditional distribution of goal for males".

Chapter 10. Inference for Regression

The descriptive analysis discussed in Chapter 2 for relations among two quantitative variables leads to formal inference. This chapter focuses on demonstrating how SPSS can be used to perform inference for **simple linear regression**.

Example 10.1
(IPS Ex. 10.1)

In practice, fat content is found by measuring body density, the weight per unit volume of the body. High fat content corresponds to low body density. Body density, denoted DEN, is hard to measure directly — the standard method requires that subjects be weighed underwater. For this reason, scientists have sought variables that are easier to measure and that can be used to predict body density. Research suggests that *skinfold thickness* can accurately predict body density. To measure skinfold thickness, pinch a fold of skin between calipers at four body locations to determine the thickness, and add the four thicknesses. LSKIN is the natural logarithm of the sum of the skinfold measures. It is of interest to determine whether a linear relationship exists between DEN and LSKIN and to model any such relationship. A scatterplot of the variables DEN and LSKIN confirms that a linear relationship exists (Figure 10.1). Directions on how to construct a scatterplot appear in Section 2.1. It is now of interest to model this relationship and perform some inference procedures.

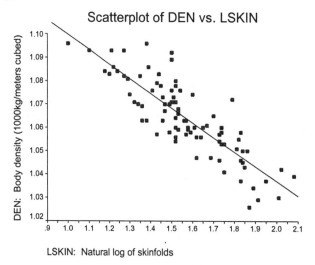

Figure 10.1

To obtain the correlation, the coefficients for the least squares regression line, test statistics, p-values, and the 95% confidence intervals, follow these steps.

1. Click **Analyze**, click **Regression**, and then click **Linear**. The SPSS window in Figure 2.3 appears with the variables *lskin* and *den*.
2. Click *lskin*, then click ▶ to move *lskin* into the "Independent(s)" box.
3. Click *den*, then click ▶ to move *den* into the "Dependent" box.
4. Click the **Statistics** box. Select **Estimates**, **Confidence intervals**, and **Model fit**. After the selections have been made, the window looks like Figure 10.2. The result of these selections appears in Table 10.2.
5. If you are interested in a normal probability plot for the residuals, click the **Plots** box (Figure 10.3 appears). Select the **Normal probability plot** option in the "Standardized Residual Plots" box. Click **Continue**.
6. If you are interested in saving predicted values, residuals, distance measures, influential statistics, and prediction intervals, click the **Save** box. Check the desired options. After the selections have been made, the window looks like Figure 10.4. Click **Continue**.
7. Click **OK**.

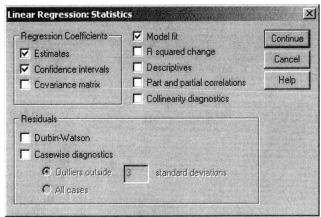

Figure 10.2

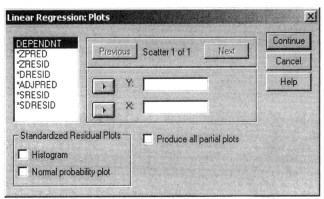

Figure 10.3

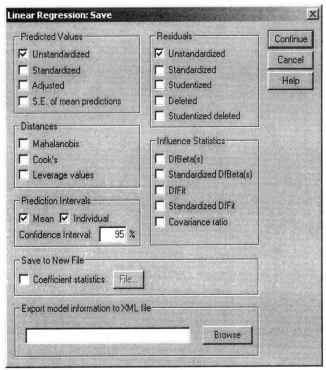

Figure 10.4

Tables 10.1 and 10.2 and Figure 10.5 (slightly edited) are part of the resulting SPSS output.

Model Summary[b]

Model	R	R Square	Adjusted R Square	Std. Error of the Estimate
1	.849[a]	.720	.717	.0085390

a. Predictors: (Constant), LSKIN
b. Dependent Variable: DEN

Table 10.1

Coefficients[a]

Model		Unstandardized Coefficients		Standardized Coefficients	t	Sig.	95% Confidence Interval for B	
		B	Std. Error	Beta			Lower Bound	Upper Bound
1	(Constant)	1.163	.007		177.296	.000	1.150	1.176
	LSKIN	-6.3E-02	.004	-.849	-15.228	.000	-.071	-.055

a. Dependent Variable: DEN

Table 10.2

Normal P-P Plot of Regression Standardized Residual

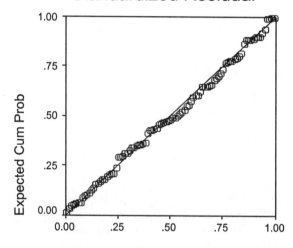

Dependent Variable: DEN

Figure 10.5

Example 10.1 (cont.)

Table 10.1 shows the correlation, r^2, and the mean squared error, while Table 10.2 shows the y-intercept and the slope. The correlation of $r = 0.849$ indicates that a strong, positive, linear association exists between DEN and LSKIN. 72% (r^2) of the variation in the skin density values is explained by the regression of skin density on the log of skin thicknesses. In addition, the equation of the least squares regression line is

$$LSKIN = 1.163 - 0.0631 DEN.$$

The estimated standard deviation of the model (often referred to as the mean squared error, MSE) is 0.008539.

Now that a fitted model has been developed, the residuals of the model should be examined. Figure 10.5 is a normal probability plot of the standardized residuals. Since the plotted values are fairly linear, the assumption of normality seems reasonable. In addition, we can examine whether the residuals display any systematic pattern when plotted against other variables. Recall that the unstandardized residuals (and various other values) were saved into the Data Editor in step 5. Figure 10.6 displays the Data Editor containing the values that were saved, and the variable **res_1** contains the residuals. Figure 10.9 is a scatterplot of **res_1** and **lskin**. No unusual patterns or values are observed. (Directions for making this scatterplot are below and include Figures 10.7 and 10.8.)

Higher body fat reduces body density and increases skinfold thickness, so we might expect a negative association. The hypotheses appropriate for testing this conjecture are $H_0: \beta_1 = 0$ and $H_a: \beta_1 < 0$. According to Table 10.2, the computed test statistic $t = -15.228$ and the P-value < 0.001. This indicates that there is sufficient evidence to conclude that a strong negative relationship exists between DEN and LSKIN. [Note that, in general, the P-value is computed assuming a two-tailed alternative, and the one-tailed P-value is half of the two-sided P-value.] Table 10.2 also shows that a 95% confidence interval for β_1 is $(-0.071, -0.055)$. This indicates that an increase of 1 in the logarithm of the skinfold thickness measure is associated with a decrease of body density between 55 and 71 kilograms per cubic meter.

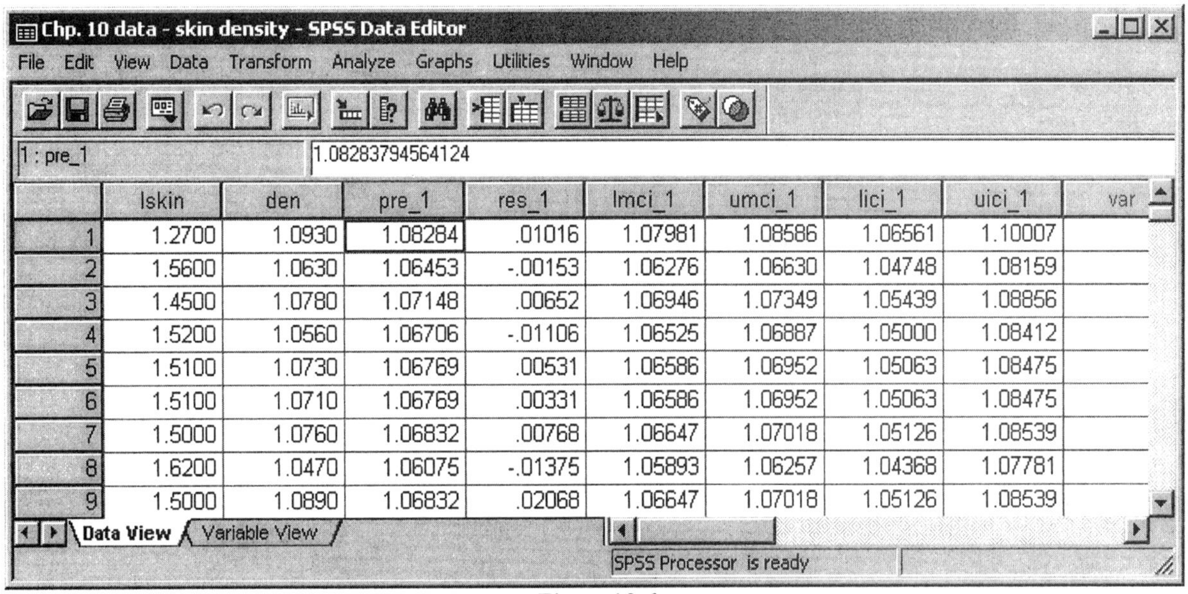

	lskin	den	pre_1	res_1	lmci_1	umci_1	lici_1	uici_1	var
1	1.2700	1.0930	1.08284	.01016	1.07981	1.08586	1.06561	1.10007	
2	1.5600	1.0630	1.06453	-.00153	1.06276	1.06630	1.04748	1.08159	
3	1.4500	1.0780	1.07148	.00652	1.06946	1.07349	1.05439	1.08856	
4	1.5200	1.0560	1.06706	-.01106	1.06525	1.06887	1.05000	1.08412	
5	1.5100	1.0730	1.06769	.00531	1.06586	1.06952	1.05063	1.08475	
6	1.5100	1.0710	1.06769	.00331	1.06586	1.06952	1.05063	1.08475	
7	1.5000	1.0760	1.06832	.00768	1.06647	1.07018	1.05126	1.08539	
8	1.6200	1.0470	1.06075	-.01375	1.05893	1.06257	1.04368	1.07781	
9	1.5000	1.0890	1.06832	.02068	1.06647	1.07018	1.05126	1.08539	

Figure 10.6

To generate a scatterplot of the residuals with $y = 0$ reference line, follow these steps.

1. Construct a scatterplot as in Section 2.1, with *lskin* in the "X Axis" box and *res_1* in the "Y Axis" box.
2. When the scatterplot appears in the Output Window, double-click on it to edit it in the Chart Editor. Make any edits that are desired to the scales, titles, footnotes, etc.
3. It is desirable to have a reference line at $y = 0$ to better detect whether or not the residuals have a pattern. To create this line, click **Chart** and then click **Reference Line**, then proceed as instructed in Section 2.2 of this manual. The result is Figure 10.7.

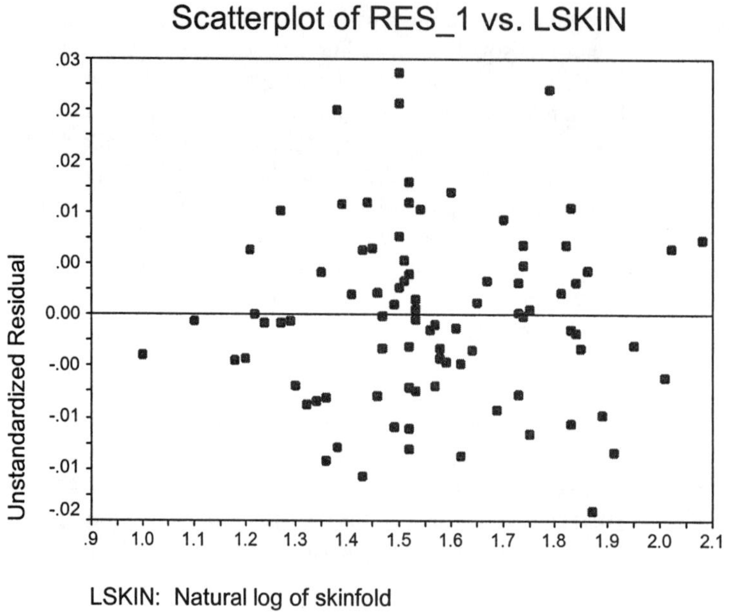

Figure 10.7

Example 10.1 (cont.)

It is also of interest to construct confidence intervals for the mean response and prediction intervals for a future observation. Figure 10.6 displays 95% confidence intervals for the mean response for each observation in the data set and 95% prediction intervals for future observations equal to each observation in the data set. The variables (*lmci_1*, *umci_1*) represent the 95% confidence interval, and the variables (*lici_1*, *uici_1*) represent the 95% prediction interval. In addition, the 95% confidence limits for the mean response (the inner bands) and the 95% prediction limits for the individual responses (the outer bands) for the body density example are displayed in Figure 10.9.

To generate a scatterplot with the regression line, the confidence limits, and the prediction limits, follow these steps.

1. Construct a scatterplot as in Section 2.1, with *lskin* in the "X Axis" box and *den* in the "Y Axis" box.
2. When the scatterplot appears in the Output Window, double-click on it to edit it in the Chart Editor. Make any edits that are desired to the scales, titles, footnotes, etc.
3. Click **Chart**, click **Options**, and then click **Total**, which is found in the "Fit Line" box. (See Figure 2.4.)
4. Click **Fit Options**. Figure 10.8 appears.
5. The Fit Method defaults to the desired "Linear regression" option. If confidence limits and prediction limits are desired, click on the options in the "Regression Prediction Line(s)" box.
6. Click **Continue**, then **OK**. Figure 10.9 is the resulting scatterplot.

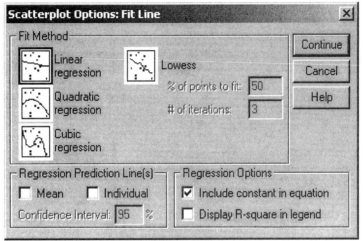

Figure 10.8

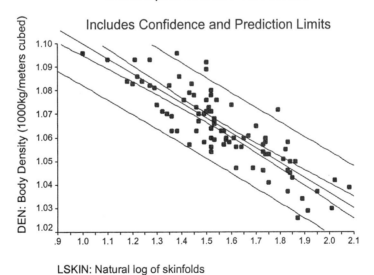

Figure 10.9

Chapter 11. Multiple Regression

This chapter focuses on using SPSS to perform **multiple regression**. Multiple regression is a technique that is used when two or more explanatory variables are used to predict a dependent variable.

Example 11.1
(IPS Ex. 11.1)

The purpose of a study conducted at a large university was to attempt to predict the success of first-year computer science majors. One measure of success was the cumulative grade point average (GPA) after three semesters. Among the explanatory variables recorded at the time the students enrolled in the university were their SAT mathematics score (SATM), their SAT verbal score (SATV), and their average high school grades in mathematics (HSM), science (HSS), and English (HSE). Note that the high school grades are coded on a scale from 1 to 10, where A = 10, A– = 9, B+ = 8, and so on, but the GPA is coded on the traditional 4-point scale. There were 224 subjects in the study. An excerpt of the complete data set is given in Table 11.1. One objective is to examine the relationships between all the pairs of variables *gpa*, *satm*, *satv*, *hsm*, *hss*, and *hse* by computing all the pairwise correlations.

Table 11.1 GPA Data Set

Sub	GPA	SATM	SATV	HSM	HSS	HSE
001	3.32	670	600	10	10	10
002	2.26	700	640	6	8	5
.	.	.	.	.	.	.
.	.	.	.	.	.	.
223	2.59	630	470	5	4	7
224	2.28	559	488	9	8	9

To obtain all of the pairwise correlations, follow these steps.
1. Click **Analyze**, click **Correlate**, and then click **Bivariate**.
2. Click *gpa*, *hsm*, *hss*, *hse*, *satm*, and *satv* then click ▶ to move *gpa*, *hsm*, *hss*, *hse*, *satm*, and *satv* into the "Variables" box. This results in Figure 11.1.
3. Click **OK**.

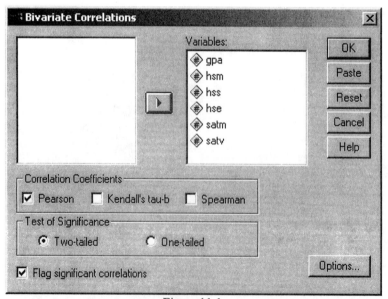

Figure 11.1

Table 11.2 is the resulting SPSS output and includes each pairwise correlation with its respective *p*-value.

Correlations

		GPA	HSM	HSS	HSE	SATM	SATV
GPA	Pearson Correlation	1	.436**	.329**	.289**	.252**	.114
	Sig. (2-tailed)	.	.000	.000	.000	.000	.087
	N	224	224	224	224	224	224
HSM	Pearson Correlation	.436**	1	.576**	.447**	.454**	.221**
	Sig. (2-tailed)	.000	.	.000	.000	.000	.001
	N	224	224	224	224	224	224
HSS	Pearson Correlation	.329**	.576**	1	.579**	.240**	.262**
	Sig. (2-tailed)	.000	.000	.	.000	.000	.000
	N	224	224	224	224	224	224
HSE	Pearson Correlation	.289**	.447**	.579**	1	.108	.244**
	Sig. (2-tailed)	.000	.000	.000	.	.106	.000
	N	224	224	224	224	224	224
SATM	Pearson Correlation	.252**	.454**	.240**	.108	1	.464**
	Sig. (2-tailed)	.000	.000	.000	.106	.	.000
	N	224	224	224	224	224	224
SATV	Pearson Correlation	.114	.221**	.262**	.244**	.464**	1
	Sig. (2-tailed)	.087	.001	.000	.000	.000	.
	N	224	224	224	224	224	224

**. Correlation is significant at the 0.01 level (2-tailed).

Table 11.2

Example 11.1 (cont.) All the high school grades have higher correlations with GPA than SAT scores. For example, the correlation between GPA and HSE is 0.289 with a *p*-value of 0.000, but the correlation between GPA and SATV is 0.114 with a *p*-value of 0.087. By examining the table of correlations, we can ascertain the strength of the relationships between the various pairwise variables. Since all the high school grades have higher correlations with GPA than SAT scores, it is reasonable to model the response variable GPA using the explanatory variables HSM, HSS, and HSE.

To obtain the multiple regression equation, follow these steps.
1. Click **Statistics**, click **Regression**, and then click **Linear**.
2. Click *gpa*, then click ▸ to move *gpa* into the "Dependent" box.
3. Click *hsm*, *hss*, and *hse*, then click ▸ to move *hsm*, *hss*, and *hse* into the "Independent(s)" box. This results in Figure 11.2.
4. If you are interested in a normal probability plot for the residuals, click the **Plots** box. Select the **Normal probability plot** option in the "Standardized Residual Plots" box. Click **Continue**.
5. If you are interested in saving predicted values, residuals, distance measures, influential statistics, and prediction intervals, click the **Save** box. Check the desired options and click **Continue**, then **OK**.

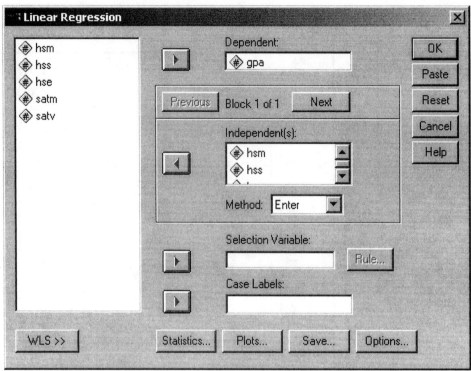

Figure 11.2

Tables 11.3 through 11.5 and Figure 11.3 are part of the resulting SPSS output.

Model Summary[b]

Model	R	R Square	Adjusted R Square	Std. Error of the Estimate
1	.452[a]	.205	.194	.69984

a. Predictors: (Constant), HSE, HSM, HSS

b. Dependent Variable: GPA

Table 11.3

ANOVA[b]

Model		Sum of Squares	df	Mean Square	F	Sig.
1	Regression	27.712	3	9.237	18.861	.000[a]
	Residual	107.750	220	.490		
	Total	135.463	223			

a. Predictors: (Constant), HSE, HSM, HSS

b. Dependent Variable: GPA

Table 11.4

Coefficients[a]

Model		Unstandardized Coefficients		Standardized Coefficients	t	Sig.
		B	Std. Error	Beta		
1	(Constant)	.590	.294		2.005	.046
	HSM	.169	.035	.354	4.749	.000
	HSS	3.432E-02	.038	.075	.914	.362
	HSE	4.510E-02	.039	.087	1.166	.245

a. Dependent Variable: GPA

Table 11.5

Example 11.1 (cont.) Initially, we want to determine whether any of the variables are useful for predicting GPA. Specifically, the null and alternative hypotheses are $H_0: \beta_1 = \beta_2 = \beta_3 = 0$ and H_a: not all of the β_i are equal to 0. According to the ANOVA table in Table 11.4, the F statistic is 18.861 with a P-value < 0.001. Therefore, at least one of the regression coefficients differs from 0 in the fitted regression model

$$GPA = 0.590 + 0.169\,HSM + 0.0343\,HSS + 0.0451\,HSE.$$

Note that the coefficients for this model came from Table 11.5. In addition, the value for R^2 is 0.205, which is to say that approximately 20% of the variation in GPA is explained by high school grades in math, science, and English (found in Table 11.3).

The next step in the analysis is to determine which variables are useful in predicting GPA in the presence of the other variables. The variables and respective P-values from Table 11.5 are: HSM (P-value < 0.001), HSS (P-value $= 0.362$), and HSE (P-value $= 0.245$). Clearly, HSM is the most significant variable in the presence of the two remaining variables. Further analysis is necessary to determine the best model for predicting GPA.

The normal distribution of the errors is necessary for this type of analysis. This assumption can be checked with a normal probability plot. Since the normal probability plot in Figure 11.3 appears fairly linear, the assumption of normality seems reasonable.

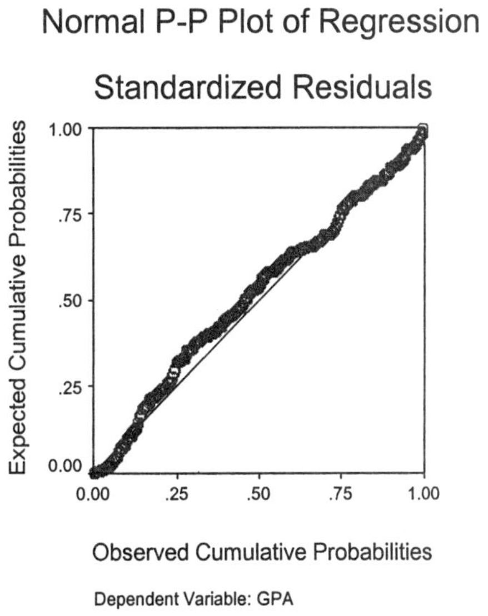

Normal P-P Plot of Regression Standardized Residuals

Figure 11.3

Chapter 12. One-Way Analysis of Variance

This chapter describes how to perform **one-way ANOVA** using SPSS for determining whether the means from several populations differ. Specifically, the null and alternate hypotheses for one-way ANOVA are $H_0: \mu_1 = \mu_2 = \ldots = \mu_I$ and H_a: not all of the μ_i are equal.

Example 12.1
(IPS Ex. 12.6)

A study of reading comprehension in children compared three methods of instruction. As is common in such studies, several pretest variables were measured before any instruction was given. One purpose of the pretest was to see if the three groups of children were similar in their comprehension skills. One of the pretest variables called SCORE was an "intruded sentences" measure, which measures one type of reading comprehension skill. The data for 22 subjects in each group are given in Table 12.1. The three groups, which are Basal, DRTA, and Strategies (denoted in the table by Strat), represent the method of instruction the student subject will receive.

Note that when entering this data set into SPSS, the variable GROUP must be numerically coded (labels may also be utilized). In this example, codes and labels are as follows: 1 = Basal, 2 = DRTA, and 3 = Strategies.

Table 12.1 Reading Scores

Sub	GROUP	SCORE	COMP	Sub	GROUP	SCORE	COMP	Sub	GROUP	SCORE	COMP
1	Basal	4	41	23	DRTA	7	31	45	Strat	11	53
2	Basal	6	41	24	DRTA	7	40	46	Strat	7	47
3	Basal	9	43	25	DRTA	12	48	47	Strat	4	41
4	Basal	12	46	26	DRTA	10	30	48	Strat	7	49
5	Basal	16	46	27	DRTA	16	42	49	Strat	7	43
6	Basal	15	45	28	DRTA	15	48	50	Strat	6	45
7	Basal	14	45	29	DRTA	9	49	51	Strat	11	50
8	Basal	12	32	30	DRTA	8	53	52	Strat	14	48
9	Basal	12	33	31	DRTA	13	48	53	Strat	13	49
10	Basal	8	39	32	DRTA	12	43	54	Strat	9	42
11	Basal	13	42	33	DRTA	7	55	55	Strat	12	38
12	Basal	9	45	34	DRTA	6	55	56	Strat	13	42
13	Basal	12	39	35	DRTA	8	57	57	Strat	4	34
14	Basal	12	44	36	DRTA	9	53	58	Strat	13	48
15	Basal	12	36	37	DRTA	9	37	59	Strat	6	51
16	Basal	10	49	38	DRTA	8	50	60	Strat	12	33
17	Basal	8	40	39	DRTA	9	54	61	Strat	6	44
18	Basal	12	35	40	DRTA	13	41	62	Strat	11	48
19	Basal	11	36	41	DRTA	10	49	63	Strat	14	49
20	Basal	8	40	42	DRTA	8	47	64	Strat	8	33
21	Basal	7	54	43	DRTA	8	49	65	Strat	5	45
22	Basal	9	32	44	DRTA	10	49	66	Strat	8	42

Example 12.1 (cont.)

Before proceeding with ANOVA, we must verify that the assumptions of (1) normally distributed data and (2) equality of standard deviations in the various groups are reasonably satisfied. Because the data appear reasonably normal (which can be demonstrated using normal quantile plots for the three groups of data) and the assumption of equal standard deviations is reasonably satisfied, we can proceed with ANOVA.

Initially, we want to analyze the pretest scores and determine whether the three groups had similar population means on this measure. Specifically, the null and alternative hypotheses are $H_0: \mu_B = \mu_D = \mu_S$ and H_a: not all of the μ_i are equal.

To perform one-way ANOVA, follow these steps.
1. Click **Analyze**, click **Compare Means**, and then click **One-Way ANOVA**. The SPSS window in Figure 12.1 appears.

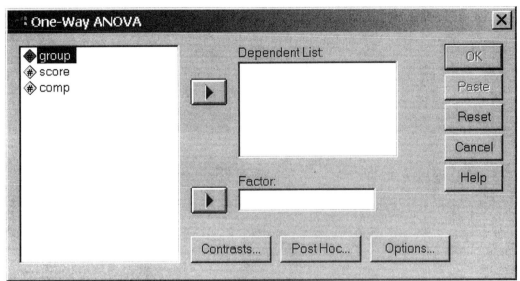

Figure 12.1

2. Click *score*, then click ▸ to move *score* into the "Dependent List" box.
3. Click *group*, then click ▸ to move *group* into the "Factor" box.
4. Click **OK**.

Table 12.2 is the resulting SPSS output.

ANOVA

SCORE

	Sum of Squares	df	Mean Square	F	Sig.
Between Groups	20.576	2	10.288	1.132	.329
Within Groups	572.455	63	9.087		
Total	593.030	65			

Table 12.2

Example 12.1 (cont.) A *P*-value of 0.329 implies that there is no reason to reject H_0 that the three groups have equal population means on this measure. This was the desired outcome. We now turn to the response variable, a measure of reading comprehension called *comp*, which was measured by a test taken after the instruction was completed. Initially, we are interested in comparing several summary statistics for *comp* in the three groups of the study.

To obtain descriptive statistics for a response variable for each category of the grouping variable, follow these steps.
1. Click **Analyze**, click **Compare Means**, and then click **Means**. The SPSS window in Figure 12.2 appears.
2. Click *comp*, then click ▸ to move *comp* into the "Dependent List" box.
3. Click *group*, then click ▸ to move *group* into the "Independent List" box.

4. If you are interested in choosing summary statistics options, click the **Options** box. Choose the desired summary statistics, and click **Continue.**
5. Click **OK**.

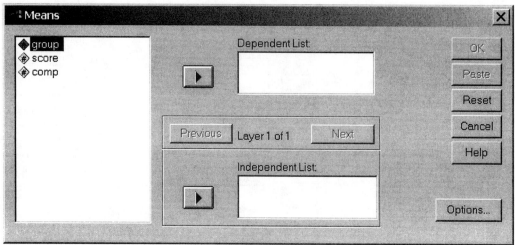

Figure 12.2

Table 12.3 is the resulting SPSS output.

Report

COMP

GROUP	Mean	N	Std. Deviation
Basal	41.0455	22	5.63558
DRTA	46.7273	22	7.38842
Strategies	44.2727	22	5.76675
Total	44.0152	66	6.64366

Table 12.3

Example 12.1 (cont.)

Given the preceding results, we are interested in analyzing the *comp* scores to determine whether the three groups have similar population means for this measure. Specifically, the hypotheses of interest are H_0: $\mu_B = \mu_D = \mu_S$ and H_a: not all of the μ_i are equal. In addition, the instruction for the Basal group was the standard method commonly used in the schools, and the DRTA and Strategies groups received innovative methods of teaching that are designed to improve reading comprehension. The researcher is interested in two primary questions: (1) whether the two new methods are better than the standard method and (2) whether the two new methods differ. These questions can be formulated using the following hypotheses:

$$H_{01}: 0.5(\mu_D + \mu_S) = \mu_B \quad \text{versus} \quad H_{a1}: 0.5(\mu_D + \mu_S) > \mu_B$$

and

$$H_{02}: \mu_D = \mu_S \quad \text{versus} \quad H_{a2}: \mu_D \neq \mu_S.$$

These null hypotheses can be formulated using the two contrasts:

$$\psi_1 = (-1)\mu_B + (0.5)\mu_D + (0.5)\mu_S$$

and

$$\psi_2 = (0)\mu_B + (1)\mu_D + (-1)\mu_S.$$

To perform a one-way ANOVA with planned contrasts, follow these steps.
1. Click **Analyze**, click **Compare Means**, and then click **One-Way ANOVA**.
2. Click *comp*, then click ▸ to move *comp* into the "Dependent List" box.
3. Click *group*, then click ▸ to move *group* into the "Factor" box.
4. If you are interested in using contrasts, click the **Contrasts** box. Enter the coefficients of your first contrast. After each coefficient, click **Add**. After each contrast is complete, click **Next**. The SPSS window in Figure 12.3 appears. After the last contrast is entered, click **Continue.**
5. Click **OK**.

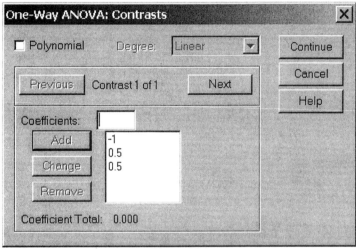

Figure 12.3

Tables 12.4 through 12.6 are the resulting SPSS output.

ANOVA

COMP

	Sum of Squares	df	Mean Square	F	Sig.
Between Groups	357.303	2	178.652	4.481	.015
Within Groups	2511.682	63	39.868		
Total	2868.985	65			

Table 12.4

Contrast Coefficients

	GROUP		
Contrast	Basal	DRTA	Strategies
1	-1	.5	.5
2	0	1	-1

Table 12.5

Contrast Tests

		Contrast	Value of Contrast	Std. Error	t	df	Sig. (2-tailed)
COMP	Assume equal variances	1	4.4545	1.64872	2.702	63	.009
		2	2.4545	1.90378	1.289	63	.202
	Does not assume equal variances	1	4.4545	1.56264	2.851	47.945	.006
		2	2.4545	1.99823	1.228	39.661	.227

Table 12.6

Example 12.1 (cont.) Since the P-value is 0.015 (from Table 12.4), there is sufficient evidence to reject H_0 in favor of H_a. This indicates that some μ_i differs. In addition, since the P-value in Table 12.6 for the first contrast is 0.0045 (0.009/2), there is sufficient evidence to reject H_{01} in favor of H_{a1}. This indicates that new methods produce higher scores than the old. Since the P-value for the second contrast is 0.202, there is insufficient evidence to reject H_{02} that the means for the two new methods are equal.

In many studies specific questions cannot be formulated in advance. As before, the hypotheses of interest are H_0: $\mu_B = \mu_D = \mu_S$ and H_a: not all of the μ_i are equal. In addition, if H_0 is rejected, we would like to know which means differ. Two procedures that are commonly used for multiple comparisons are the least-significant differences (LSD) method and the Bonferroni method.

The ANOVA table (which is the same as Table 12.4) indicates that some μ_i differs. The output of the multiple comparisons tables indicates significantly different means with an asterisk (*). Both the LSD and the Bonferroni methods indicate that the means for the Basal and the DRTA groups differ.

To perform a one-way ANOVA with post hoc multiple comparisons, follow these steps.
1. Click **Analyze**, click **Compare Means**, and then click **One-Way ANOVA**.
2. Click *comp*, then click ▶ to move *comp* into the "Dependent List" box.
3. Click *group*, then click ▶ to move *group* into the "Factor" box.
4. If you are interested in multiple comparisons, you can click on the **Post Hoc** box. The SPSS window in Figure 12.4 appears. Click on the desired multiple comparison test procedure, and click **Continue.**

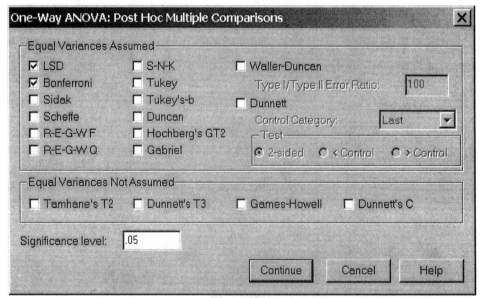

Figure 12.4

5. If you are interested in either descriptive statistics, a means plot, or Levene's Test for the Equality of Variances, you can click on the **Options** box. The "One-Way ANOVA: Options" window appears. If you are interested in descriptive statistics, click **Descriptive**. If you are interested in a means plot, click **Means plot**. If you are interested in performing Levene's Test, click **Homogeneity-of-variance**. Then click **Continue.** Figure 12.5 demonstrates the SPSS window after the means plot option has been selected.
6. Click **OK**.

For illustration purposes, SPSS output from a means plot (Figure 12.6) and both the LSD and the Bonferroni methods (Table 12.7) have been provided.

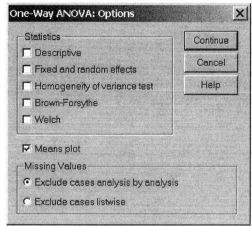

Figure 12.5

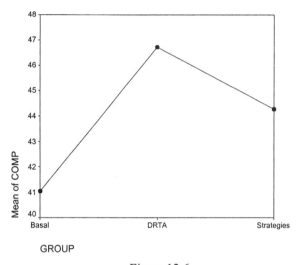

Figure 12.6

Multiple Comparisons

Dependent Variable: COMP

	(I) GROUP	(J) GROUP	Mean Difference (I-J)	Std. Error	Sig.	95% Confidence Interval	
						Lower Bound	Upper Bound
LSD	Basal	DRTA	-5.6818*	1.90378	.004	-9.4862	-1.8774
		Strategies	-3.2273	1.90378	.095	-7.0317	.5771
	DRTA	Basal	5.6818*	1.90378	.004	1.8774	9.4862
		Strategies	2.4545	1.90378	.202	-1.3498	6.2589
	Strategies	Basal	3.2273	1.90378	.095	-.5771	7.0317
		DRTA	-2.4545	1.90378	.202	-6.2589	1.3498
Bonferroni	Basal	DRTA	-5.6818*	1.90378	.012	-10.3643	-.9993
		Strategies	-3.2273	1.90378	.285	-7.9098	1.4552
	DRTA	Basal	5.6818*	1.90378	.012	.9993	10.3643
		Strategies	2.4545	1.90378	.606	-2.2280	7.1370
	Strategies	Basal	3.2273	1.90378	.285	-1.4552	7.9098
		DRTA	-2.4545	1.90378	.606	-7.1370	2.2280

*. The mean difference is significant at the .05 level.

Table 12.7

Chapter 13. Two-Way Analysis of Variance

This chapter describes how to perform **two-way ANOVA** using SPSS for determining whether the means from several populations that are classified in two ways differ. Recall that one-way designs vary a single factor and hold other factors fixed. In a two-way ANOVA model there are two factors that vary. The researcher can investigate the effects of each main factor and the interaction between the factors.

Example 13.1
(IPS Ex. 13.8)

A study of cardiovascular risk factors compared men and women runners who averaged at least 15 miles per week with a control group of men and women described as "generally sedentary." The design is a 2 × 2 ANOVA with the factors group and gender. There were 200 subjects in each of the four combinations. The response variable of interest was the heart rate after 6 minutes of exercise on a treadmill. An excerpt of the complete dataset is given in Table 13.1. In this example, GROUP and GENDER must be numerically coded as follows: 0 = Control, 1 = Runner, 0 = Female, and 1 = Male.

Table 13.1 Heart Rate

Sub	GROUP	GENDER	RATE
001	Control	Female	159
002	Control	Female	183
003	Control	Female	140
.	.	.	.
.	.	.	.
798	Runner	Male	112
799	Runner	Male	97
800	Runner	Male	89

Before proceeding with ANOVA, we must verify that the assumptions of (1) normally distributed data and (2) equality of standard deviations in the various groups are reasonably satisfied. Because the data appear reasonably normal (which can be demonstrated using normal quantile plots for the four groups of data) and the assumption of equal standard deviations is reasonably satisfied, we can proceed with ANOVA.

Initially, we are interested in comparing several summary statistics for each of the four combinations of the factors of the study. Table 13.2 is the result of SPSS analysis (the instructions will be given later in this chapter).

Inspection of Table 13.2 indicates that there may be a difference in the means between the groups and the genders. Therefore, we are interested in analyzing data to determine whether the factors influence the mean heart rate.

Descriptive Statistics

Dependent Variable: Heart Rate

GROUP	GENDER	Mean	Std. Deviation	N
Control	Female	148.0000	16.27095	200
	Male	130.0000	17.10035	200
	Total	139.0000	18.94961	400
Runner	Female	115.9850	15.97154	200
	Male	103.9750	12.49942	200
	Total	109.9800	15.53376	400
Total	Female	131.9925	22.71889	400
	Male	116.9875	19.83724	400
	Total	124.4900	22.59692	800

Table 13.2

To perform this two-way ANOVA, follow these steps.
1. Click **Analyze**, click **General Linear Model**, and then click **Univariate**. The SPSS window in Figure 13.1 appears.

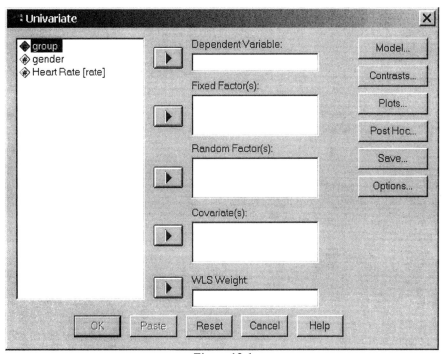

Figure 13.1

2. Click *rate*, then click ▸ to move *rate* into the "Dependent Variable" box.
3. Click *group* and *gender*, then click ▸ to move *group* and *gender* into the "Fixed Factor(s)" box.
4. The default model for SPSS is the full factorial model (i.e., the model that has all the main effects and interaction effects) computed using type III sums of squares. If you are interested in specifying a custom model, click the **Model** box. Click the **Custom** circle and choose the desired terms for the model (main effects, interaction terms, etc.). In addition, you can choose the desired sum of squares for the model (type I, II, III, or IV), and click **Continue**. In our example, we selected the type I sums of squares. The SPSS window found in Figure 13.2 appears.

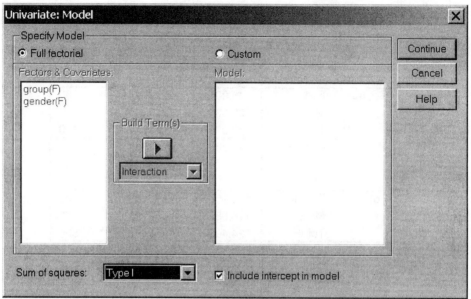

Figure 13.2

5. If you are interested in using contrasts, click the **Contrasts** box. Choose the desired type of contrast. After the selection of the contrasts are complete, click **Continue.**
6. If you are interested in constructing a means plot, click the **Plots** box. The SPSS window in Figure 13.3 appears. Click *group*, then click ▸ to move *group* into the "Horizontal Axis" box. Click *gender*, then click ▸ to move *gender* into the "Separate Lines" box. Click **Add**, and then click **Continue**.

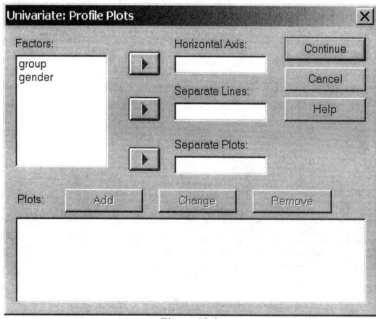

Figure 13.3

7. If you are interested in comparing several summary statistics for each of the four combinations of the factors of the study, click the **Options** box. The SPSS window in Figure 13.4 appears. Click *group* and *gender*, then click ▸ to move *group* and *gender* into the "Display Means" box. Click the **Descriptive Statistics** box, and then click **Continue**.
8. Click **OK**.

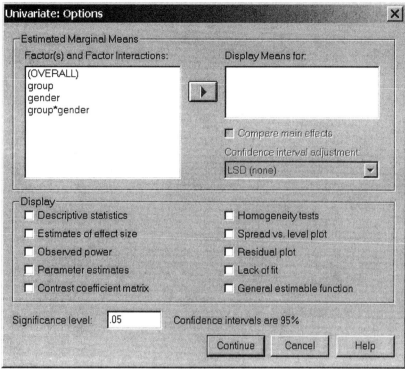

Figure 13.4

Tables 13.3 and 13.4 and Figure 13.5 are the resulting SPSS output. In addition, step 7 yields the table of descriptive statistics that was displayed in Table 13.2.

Between-Subjects Factors

		Value Label	N
GROUP	0	Control	400
	1	Runner	400
GENDER	0	Female	400
	1	Male	400

Table 13.3

Tests of Between-Subjects Effects

Dependent Variable: Heart Rate

Source	Type I Sum of Squares	df	Mean Square	F	Sig.
Corrected Model	215256.09[a]	3	71752.030	296.345	.000
Intercept	12398208	1	12398208.08	51206.26	.000
GROUP	168432.08	1	168432.080	695.647	.000
GENDER	45030.005	1	45030.005	185.980	.000
GROUP * GENDER	1794.005	1	1794.005	7.409	.007
Error	192729.83	796	242.123		
Total	12806194	800			
Corrected Total	407985.92	799			

a. R Squared = .528 (Adjusted R Squared = .526)

Table 13.4

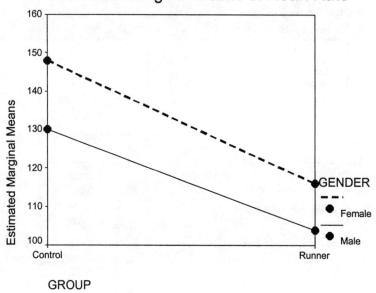

Figure 13.5

Example 13.1 (cont.)

All three effects (***group***, ***gender***, and the interaction of ***group*gender***) are statistically significant (see Table 13.4). To interpret these results, we can examine the plot of means on Figure 13.5. The significance of the main effect for ***group*** is demonstrated by the fact that the mean heart rate for the control group exceeds the mean heart rate for the runner group for both genders. The significance of the main effect of ***gender*** is demonstrated by the fact that the mean heart rate for females exceeds that of males for each group. The significance of the interaction of ***group*gender*** is demonstrated by the fact that the two lines are not parallel. This indicates that the difference in the mean heart rate between the control and runner groups is greater for women (148 − 116 = 32) than for men (130 − 104 = 26). As Figure 13.5 suggests, the interaction effect is not large, but (due to the sample size) it is statistically significant.

Chapter 14. Nonparametric Tests

This chapter introduces one type of **nonparametric** procedures, tests that can replace *t* tests and one-way analysis of variance when the normality assumption for those tests is not met. When distributions are strongly skewed, the mean may not be the preferred measure of center. The focus of these tests is on medians rather than means. All three of these tests utilize ranks of the observations in calculating the test statistic.

Section 14.1. Wilcoxon Rank Sum Test

The **Wilcoxon rank sum test** is the nonparametric counterpart of the parametric independent *t* test. It is applied to situations in which the normality assumption underlying the parametric independent *t* test has been violated or questionably met. The focus of this test is on medians rather than means. An alternate form of this test, the one used by SPSS, is the Mann-Whitney *U* test.

Example 14.1
(IPS Ex. 14.6)

Food sold at outdoor fairs and festivals may be less safe than food sold in restaurants because it is prepared in temporary locations and often by volunteer help. What do people who attend fairs think about the safety of food served? One study asked this question of people at a number of fairs in the Midwest: How often do you think people become sick because of food they consume prepared at outdoor fairs and festivals? The possible responses were: 1 = very rarely, 2 = once in a while, 3 = often, 4 = more often than not, and 5 = always.

In all, 303 people answered the question. Of these, 196 were women and 107 were men. Is there good evidence that men and women differ in their perceptions about food safety at fairs? The data are presented in Table 14.1 as a two-way table of counts.

Table 14.1 Responses by Gender

Gender	Responses					Total
	1	2	3	4	5	
Female	13	108	50	23	2	196
Male	22	57	22	5	1	107
Total	35	165	72	28	3	303

We would like to know whether men or women are more concerned about food safety. Whereas a Chi-square test could be applied to answer the general question, this test ignores the ordering of the responses and so does not use all of the available information. Since the data are ordinal, a test based on ranks makes sense. One can use the Wilcoxon rank sum test for the hypotheses:

H_0: men and women do not differ in their responses
H_a: one of the two genders gives systematically larger responses than the other.

The data were entered into SPSS using ten rows and three columns with the variable names *gender* (declared numeric 8.0 with value labels 1 = Female and 2 = Male), *sick* (declared numeric 8.0 with value labels 1 = very rarely, 2 = once in a while, 3 = often, 4 = more often than not, and 5 = always), and *weight* (declared numeric 8.0) where *weight* represents the count of individuals for each gender who selected each of the five response options. It is important to note that the grouping variable (in this case *gender*) must be a numeric variable (not entered as M's and F's). Also, prior to performing any analyses, the weighting option under **Data** and **Weight Cases** was activated (see Chapter 9 for an example of how to enter these data and how to activate the weighting option).

To conduct a Wilcoxon rank sum test (or Mann-Whitney *U* test), follow these steps.
1. Click **Analyze**, click **Nonparametric Tests**, and then click **2 Independent Samples**. The "Two-Independent Samples Tests" window in Figure 14.1 appears.

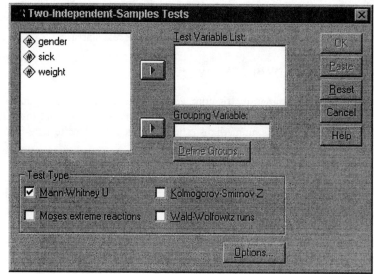

Figure 14.1

2. Click *sick*, then click ▸ to move *sick* into the "Test Variable List" box.
3. Click *gender*, then click ▸ to move *gender* into the "Grouping Variable" box.
4. Click the **Define Groups** button.
5. Type **1** in the "Group 1" box, press the **Tab** key, and type **2** in the "Group 2" box.
6. Click **Continue**.
7. The default test is the Mann-Whitney *U* test (as indicated by the ✓ in front of "Mann-Whitney *U*" in the "Test Type" area).
8. Click **OK**.

The resulting output is displayed in Tables 14.2 and 14.3.

Ranks

	GENDER	N	Mean Rank	Sum of Ranks
SICK	Female	196	163.25	31996.50
	Male	107	131.40	14059.50
	Total	303		

Table 14.2

Test Statistics[a]

	SICK
Mann-Whitney U	8281.500
Wilcoxon W	14059.500
Z	-3.334
Asymp. Sig. (2-tailed)	.001

a. Grouping Variable: GENDER

Table 14.3

Example 14.1 (cont.)

As can be seen in Table 14.2, the rank sum for men (using average ranks for ties) is $W = 14,059.5$. Since the sample size is large, the z distribution should yield a reasonable approximation of the P-value. As shown in Table 14.3, the standardized value is $z = -3.334$ with a two-sided P-value $= 0.001$. This small P-value lends strong evidence that women are more concerned than men about the safety of food served at fairs.

Section 14.2. Wilcoxon Signed Rank Test

This section will introduce the **Wilcoxon signed rank test**, the nonparametric counterpart of a paired-samples t test. It is used in situations in which there are repeated measures (the same group is assessed on the same measure on two occasions) or matched subjects (pairs of individuals are each assessed once on a measure). It is applied to situations in which the assumptions underlying the parametric t test have been violated or questionably met. The focus of this test is on medians rather than means.

Example 14.2 (IPS Ex. 14.11)

The golf scores of 12 members of a college women's golf team in two rounds of tournament play are shown in Table 14.4. A golf score is the number of strokes required to complete the course; therefore, low scores are better.

Table 14.4

Player	1	2	3	4	5	6	7	8	9	10	11	12
Round 2	94	85	89	89	81	76	107	89	87	91	88	80
Round 1	89	90	87	95	86	81	102	105	83	88	91	79
Difference	5	-5	2	-6	-5	-5	5	-16	4	3	-3	1

The data were entered into SPSS using two columns and the variable names *round_1* (declared numeric 8.0) and *round_2* (declared numeric 8.0). The variables *round_1* and *round_2* were entered into the first and second columns, respectively. The order in which the variables are entered is important as will be explained later.

Because this is a matched pairs design, inference is based on the differences between pairs. Negative differences indicate better (lower) scores on the second round. We see that 6 of the 12 golfers improved their scores. We would like to test the hypotheses that in a large population of collegiate women golfers:

H_0: scores have the same distribution in Rounds 1 and 2
H_a: scores are systematically lower or higher in Round 2.

The assessment of whether the assumption of normality has been met is based on the difference in golf scores (see Section 7.1 on the matched pairs t test for instructions in generating the difference variable). A small sample makes it difficult to assess normality adequately, but the normal quantile plot of the differences in Figure 14.2 shows some irregularity and a low outlier. As such, one should use the Wilcoxon signed rank test, which does not require normality.

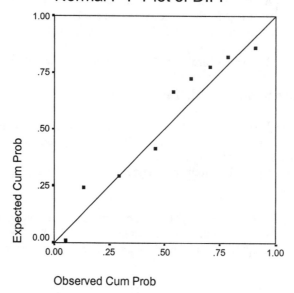

Figure 14.2

To conduct a Wilcoxon signed rank test, follow these steps.
1. Click **Analyze**, click **Nonparametric Tests**, and then click **2 Related Samples**. The "Two-Related Samples Tests" window shown in Figure 14.3 appears.
2. Click *round_1* and it appears after "Variable 1" in the "Current Selections" box.

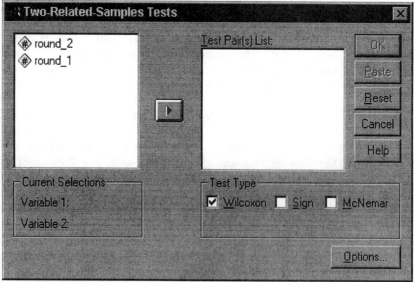

Figure 14.3

3. Click *round_2* and it appears after "Variable 2" in the "Current Selections" box.
4. Click ▸ to move the variables into the "Test Pair(s) List" box (it will read "*round_1 – round_2*").
5. The default test is the Wilcoxon signed rank test (as indicated by the ✔ in front of "Wilcoxon" in the "Test Type" box).
6. Click **OK**.

124

The resulting SPSS output is displayed in Tables 14.5 and 14.6.

Ranks

		N	Mean Rank	Sum of Ranks
ROUND_2 - ROUND_1	Negative Ranks	6[a]	8.42	50.50
	Positive Ranks	6[b]	4.58	27.50
	Ties	0[c]		
	Total	12		

a. ROUND_2 < ROUND_1
b. ROUND_2 > ROUND_1
c. ROUND_1 = ROUND_2

Table 14.5

Test Statistics[b]

	ROUND_2 - ROUND_1
Z	-.910[a]
Asymp. Sig. (2-tailed)	.363

a. Based on positive ranks.
b. Wilcoxon Signed Ranks Test

Table 14.6

Example 14.2 (cont.)

First, notice that the difference in Table 14.5 reads "ROUND_2 – ROUND_1", which is the reverse of the difference variable that appeared to be created in Step 4 (*round_1 – round_2*). For this test, SPSS always creates a difference score between the two named variables based on the order in which the variables are entered in the data set. The variable that appears first in the data set is always subtracted from the variable that appears later in the data set. For this problem, the variables were entered in the order of *round_1* and then *round_2*. Thus, SPSS creates a difference score of *round_2* minus *round_1*, despite the order in which the variables appear in the "Test(s) Pairs List" box in the "Two-Related Samples Tests" window. The same conclusion will be reached regardless of the order in which subtraction was done as the two-tailed *P*-value will be the same whether the difference is *round_2 – round_1* or *round_1 – round_2*. However, caution must be used in performing a directional test.

As shown in Table 14.5, the sum of the Negative Ranks (ROUND_2 < ROUND_1) is 50.5 and the sum of the Positive Ranks (ROUND_2 > ROUND_1) is 27.5. The value of 0 for Ties means that there were no pairs of scores in which the values were the same (e.g., ROUND_2 = ROUND_1 (it does not mean that there were no ties among the ranks)). The Wilcoxon signed rank statistic is the sum of the positive differences. The value is $W^+ = 27.5$. In Table 14.6, a *z* value of – 0.91 is reported that is based on the standardized sum of the positive ranks and is not adjusted for the continuity correction. The corresponding *P*-value is given as 0.363. These data give no evidence for a systematic change in scores between rounds..

Section 14.3. Kruskal-Wallis Test

The **Kruskal-Wallis test** is the nonparametric counterpart of the parametric one-way analysis of variance. It is applied to situations in which the normality assumption underlying the parametric one-way ANOVA has been violated or questionably met. The focus of this test is on medians rather than means.

Example 14.3
(IPS Ex. 14.13)

Lamb's-quarter is a common weed that interferes with the growth of corn. A researcher planted corn at the same rate in 16 small plots of ground, then randomly assigned plots to four groups. He weeded the plots by hand to allow a fixed number of Lamb's-quarters to grow in each meter of a corn row. These numbers were 0, 1, 3, and 9 in the four groups of plots. No other weeds were allowed to grow, and all plots received identical treatment except for the weeds. The yields of corn (bushels per acre) in each of the plots are shown in Table 14.7. The summary statistics for the data are shown in Table 14.8.

Table 14.7 Yields of Corn

Weeds per meter	Yield (bushels/acre)			
0	166.7	172.2	165.0	176.9
1	166.2	157.3	166.7	161.1
3	158.6	176.4	153.1	156.0
9	162.8	142.4	162.7	162.4

Table 14.8 Descriptive Statistics for Yield

Weeds per meter	n	Mean	Std. Dev.
0	4	170.200	5.422
1	4	162.825	4.469
3	4	161.025	10.493
9	4	157.575	10.118

Example 14.3 (cont.)

The sample standard deviations do not satisfy the rule of thumb from IPS for use of ANOVA, that the largest standard deviation should not exceed twice the smallest. Normal quantile plots show that outliers are present in the yields for 3 and 9 weeds per meter. These are the correct yields for their plots, so we have no justification for removing them. We may want to use a nonparametric test.

The null hypothesis H_0: yields have the same distribution in all groups is being tested against the one-sided alternative H_a: yields are systematically higher in some groups than in others.

The data were entered in two columns using the variables of *weeds* (declared numeric 8.0) and *yield* (declared numeric 8.1). It is important to note that the grouping variable (in this case *weeds*) must always be a numeric variable. To conduct a Kruskal-Wallis H test, follow these steps.

1. Click **Analyze**, click **Nonparametrics Tests**, and then click **K Independent Samples**. The SPSS window shown in Figure 14.4 will appear.
2. Click *yield*, then click ▶ to move *yield* into the "Test Variable List" box.
3. Click *weed*, then click ▶ to move *weed* into the "Grouping Variable" box.
4. Click **Define Range**.
5. Type **0** in the "Minimum" box, press the **Tab** key, and type **9** in the "Maximum" box.
6. Click **Continue**.
7. The default test is the Kruskal-Wallis H Test (as indicated by the ✓ in front of "Kruskal-Wallis H" in the "Test Type" box).
8. Click **OK**.

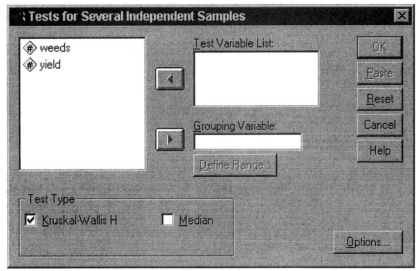

Figure 14.4

Example 14.3 (cont.) Examination of Table 14.8 suggests that an increase in weeds results in decreased yield. The output from the analysis is shown in Tables 14.9 and 14.10. As can be seen in Table 14.9, the mean rank for the group with 0 weeds per meter was 13.13, the mean rank for the group with 1 weed per meter was 8.38, the mean rank for the group with 3 weeds per meter was 6.25, and the mean rank for the group with 9 weeds per meter was 6.25. SPSS uses the Chi-square approximation to obtain a P-value = 0.134, as shown in Table 14.10. This small experiment suggests that more weeds decrease yield but does not provide convincing evidence that weeds have an effect.

Ranks

	WEEDS	N	Mean Rank
YIELD	0	4	13.13
	1	4	8.38
	3	4	6.25
	9	4	6.25
	Total	16	

Table 14.9

Test Statistics[a,b]

	YIELD
Chi-Square	5.573
df	3
Asymp. Sig.	.134

a. Kruskal Wallis Test
b. Grouping Variable: WEEDS

Table 14.10

Chapter 15. Logistic Regression

Logistic regression models are used to model the relationship between a response variable that has only two values (success and failure) and one or more explanatory variables. Specifically, logistic regression models the natural logarithm of the odds of success using the model

$$\log\left(\frac{p}{1-p}\right) = \beta_0 + \beta_1 x$$

where p is the proportion of success and x is the explanatory variable. This chapter describes how to perform **logistic regression** using SPSS.

Example 15.1
(IPS Ex. 15.7)

An experiment was designed to examine how well the insecticide rotenone kills aphids that feed on chrysanthemum plants. The explanatory variable is the log concentration (in milligrams per liter) of the insecticide. At each concentration, approximately 50 insects were exposed. Each insect was either killed or not killed. The results are given in Table 15.1.

Table 15.1

Concentration (log)	Num. of Insects	Num. Killed
0.96	50	6
1.33	48	16
1.63	46	24
2.04	49	42
2.32	50	44

Logistic regression can be used to model the relationship between the response variable, log odds of the proportion killed, and the explanatory variable, concentration. In order to enter the data for this problem into SPSS the data set should be coded as displayed in Table 15.2. Note that when entering the data, the variable *killed* should be weighted using the variable *freq*. (See Chapter 9 for directions on the weight option.)

Table 15.2

Killed	Concentration (log)	Frequency
0	0.96	44
1	0.96	6
0	1.33	32
1	1.33	16
0	1.63	22
1	1.63	24
0	2.04	7
1	2.04	42
0	2.32	6
1	2.32	44

To perform this logistic regression, follow these steps.
1. Click **Analyze**, click **Regression**, and then click **Binary Logistic**. The SPSS window in Figure 15.1 appears.
2. Click *killed*, then click ▸ to move *killed* into the "Dependent Variable" box.
3. Click *conc*, then click ▸ to move *conc* into the "Covariates" box.
4. If you are interested in a confidence interval for the odds ratio, click the **Options** box. The SPSS window in Figure 15.2 appears. Click the **CI for exp(B)** option, and then click **Continue.**
5. Click **OK**.

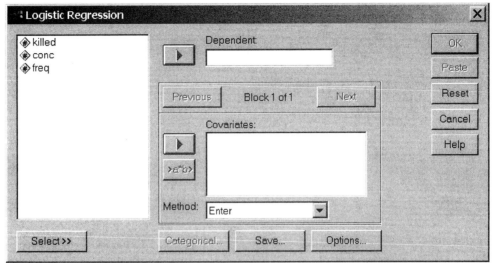

Figure 15.1

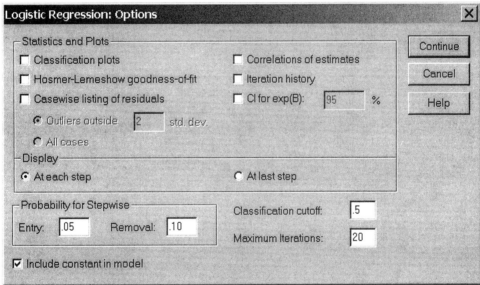

Figure 15.2

Tables 15.3 – 15.6 are part of the resulting SPSS output.

Omnibus Tests of Model Coefficients

		Chi-square	df	Sig.
Step 1	Step	95.234	1	.000
	Block	95.234	1	.000
	Model	95.234	1	.000

Table 15.3

Model Summary

Step	-2 Log likelihood	Cox & Snell R Square	Nagelkerke R Square
1	239.819	.324	.433

Table 15.4

Classification Table[a]

Observed			Predicted		
			KILLED		Percentage Correct
			0	1	
Step 1	KILLED	0	76	35	68.5
		1	22	110	83.3
	Overall Percentage				76.5

a. The cut value is .500

Table 15.5

Variables in the Equation

		B	S.E.	Wald	df	Sig.	Exp(B)	95.0% C.I.for EXP(B)	
								Lower	Upper
Step 1[a]	CONC	3.109	.388	64.233	1	.000	22.394	10.470	47.896
	Constant	-4.892	.643	57.961	1	.000	.008		

a. Variable(s) entered on step 1: CONC.

Table 15.6

Example 15.1 (cont.)

From the SPSS output in Table 15.6 one observes that the fitted model is

$$\log\left(\frac{p}{1-p}\right) = \log(odds) = -4.892 + 3.109x$$

where x is the log concentration. In addition, there is sufficient evidence to reject the null hypothesis that $\beta_1 = 0$ in favor of the alternative hypothesis that $\beta_1 \neq 0$ (since $\chi^2 = 64.23$ and P-value < 0.0005). The odds ratio, given by Exp(B), is 22.394. This indicates that an increase in 1 unit in the log concentration of the insecticide results in a 22-fold increase in the odds that an insect will be killed. The 95% confidence interval for the odds ratio is (10.470, 47.896).

Example 15.2
(IPS Ex. 15.5)

A study is conducted to examine TASTE, a measure of the quality of a particular kind of cheese. In this example, the cheese is classified as acceptable (TASTEOK=1) if TASTE is greater than or equal to 37 and unacceptable (TASTEOK=0) if TASTE is less than 37. The data set contains three explanatory variables, ACETIC, H2S, and LACTIC, which are the transformed concentrations of acetic acid, hydrogen sulfide, and lactic acid, respectively. Table 15.7 displays the data set (to preserve space TASTEOK is denoted by OK, ACETIC is denoted by ACE, and LACTIC is denoted by LAC).

Table 15.7

Sub	TASTE	OK	ACE	H2S	LAC	Sub	TASTE	OK	ACE	H2S	LAC
1	12.3	0	4.543	3.135	0.86	16	40.9	1	6.365	9.588	1.74
2	20.9	0	5.159	5.043	1.53	17	15.9	0	4.787	3.912	1.16
3	39.0	1	5.366	5.438	1.57	18	6.4	0	5.412	4.700	1.49
4	47.9	1	5.759	7.496	1.81	19	18.0	0	5.247	6.174	1.63
5	5.6	0	4.663	3.807	0.99	20	38.9	1	5.438	9.064	1.99
6	25.9	0	5.697	7.601	1.09	21	14.0	0	4.564	4.949	1.15
7	37.3	1	5.892	8.726	1.29	22	15.2	0	5.298	5.220	1.33
8	21.9	0	6.078	7.966	1.78	23	32.0	0	5.455	9.242	1.44
9	18.1	0	4.898	3.850	1.29	24	56.7	1	5.855	10.199	2.01
10	21.0	0	5.242	4.174	1.58	25	16.8	0	5.366	3.664	1.31
11	34.9	0	5.740	6.142	1.68	26	11.6	0	6.043	3.219	1.46
12	57.2	1	6.446	7.908	1.90	27	26.5	0	6.458	6.962	1.72
13	0.7	0	4.477	2.996	1.06	28	0.7	0	5.328	3.912	1.25
14	25.9	0	5.236	4.942	1.30	29	13.4	0	5.802	6.685	1.08
15	54.9	1	6.151	6.752	1.52	30	5.5	0	6.176	4.787	1.25

Logistic regression can be used to model the relationship between the response variable, log odds of the proportion of acceptable cheese, and the three explanatory variables, ACETIC, H2S, and LACTIC. Specifically, the model is

$$\log(odds) = \beta_0 + \beta_1 ACETIC + \beta_2 H2S + \beta_3 LACTIC.$$

To perform this logistic regression, follow these steps.
1. Click **Analyze**, click **Regression**, and then click **Binary Logistic**. The SPSS window in Figure 15.3 appears.

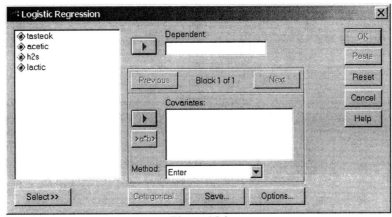

Figure 15.3

2. Click *tasteok*, then click ▸ to move *tasteok* into the "Dependent" box.
3. Click *acetic*, *h2s*, and *lactic*, then click ▸ to move *acetic*, *h2s*, and *lactic* into the "Covariates" box.

4. If you are interested in a confidence interval for the odds ratio, click the **Options** box. Click the **CI for exp(B)** option, then click **Continue**.
5. Click **OK**.

Tables 15.8 – 15.11 are part of the resulting SPSS output.

Omnibus Tests of Model Coefficients

		Chi-square	df	Sig.
Step 1	Step	16.334	3	.001
	Block	16.334	3	.001
	Model	16.334	3	.001

Table 15.8

Model Summary

Step	-2 Log likelihood	Cox & Snell R Square	Nagelkerke R Square
1	18.461	.420	.612

Table 15.9

Classification Table[a]

			Predicted		
			TASTEOK		Percentage Correct
	Observed		0	1	
Step 1	TASTEOK	0	19	3	86.4
		1	3	5	62.5
	Overall Percentage				80.0

a. The cut value is .500

Table 15.10

Variables in the Equation

		B	S.E.	Wald	df	Sig.	Exp(B)
Step 1[a]	ACETIC	.584	1.544	.143	1	.705	1.794
	H2S	.685	.404	2.873	1	.090	1.983
	LACTIC	3.468	2.650	1.713	1	.191	32.084
	Constant	-14.260	8.287	2.961	1	.085	.000

a. Variable(s) entered on step 1: ACETIC, H2S, LACTIC.

Table 15.11

Example 15.2 (cont.)

From the output the fitted model is
$$\log(odds) = -14.26 + 0.58\,ACETIC + 0.68\,H2S + 3.47\,LACTIC.$$

Initially, we examine the null and alternative hypotheses H_0: $\beta_1 = \beta_2 = \beta_3 = 0$ and H_a: some β_i differs from 0. This hypothesis is tested using the Chi-square statistic for the model. Since the test statistic $\chi^2 = 16.33$ (df = 3) and the P-value < 0.001, we reject H_0 and conclude that at least one of the explanatory variables can be used to predict the odds that cheese is acceptable. Further analyses are necessary to determine the explanatory variable(s) that results in the best model. These analyses are assigned in IPS homework exercises 15.9, 15.10, and 15.11.

Index

Adding an Observation or a Variable, 16
Analysis of Variance (One-way), 110
Analysis of Variance (Two-way), 116
ASCII Data Set, 9
Bar Charts
    Based on Conditional Distribution, 96
    Editing, 31
    For Grouped Data, 37
    For Ungrouped Data, 29
Binomial Data, Generating, 73
Binomial Probability Distribution, 70
Bonferroni Multiple Comparisons, 114
Boxplot, 49,
    Side-by-Side, 53
Chi-Square Test, 95
Color, Changing on a Chart, 31
Column, Changing Width, 4
Comparing Distributions, 51
Confidence Intervals, 77
    Difference in Means Based on t, 89
    Mean Difference Based on t, 84
    Mean Response, Regression, 104
    Odds Ratio, 128, 130
    One-Sample t, 80
Contingency Tables (see Two-Way Tables)
Contrasts, Planned with Analysis of Variance, 113
Converting an ASCII Data File into a Excel File, 9
Copying from SPSS into Microsoft Word, 24
Correlation, 60, 102, 107
Creating Variables, 54, 55, 56, 58, 70, 82, 84
Crosstabs, 93, 94, 96
Decimal Places, Changing, 5
Defining a Variable, 2
Deleting
    Observation or a Variable, 17
    Output, 22
Descriptive Statistics, 48
    Comparing Across Distributions, 51
Editing
    Bar Charts, 31
    Histograms, 43
    Pie Charts, 35
Entering Data, 2
    Categorical Data, 3
    Numerical Data, 5
Excel
    Converting an ASCII Data Set Into, 9
    Opening an Existing File in SPSS, 14
    Saving in a Format Usable by SPSS, 13

Filter Variable, 19, 68, 97
Frequency Table, Ungrouped Data, 28
Generating
    Binomial Data, 73
    Normal Data, 74
Help in SPSS, 25
Histograms, 42
Kruskal-Wallis Test, 126
Interaction Plot of Means, Two-Way ANOVA, 118, 120
Least-Squares Regression Line, 61, 62
Legend, Removing on a Chart, 45
Levene's Test, 89
Linear Transformations, 54
Logistic Regression, 128
LSD Multiple Comparisons, 114
Mann-Whitney U Test (see Wilcoxon Rank Sum Test)
Matched-Pairs t-test, 83
Mean, 48
Median, 48
Minimum and Maximum, 48
Multiple Linear Regression, 106
Navigating in SPSS, 1
Nonparametric Tests, 121
    Chi-Square, 95
    Kruskal-Wallis, 126
    Sign Test, 86
    Wilcoxon Rank Sum, 121
    Wilcoxon Signed Rank, 123
Normal Data, Generating, 74
Normal Distributions, 55, 72
Normal Quantile Plots, 56, 76, 102, 109. 110
Numbers, Adding to a Bar Chart, 31
Numeric Variables, 5
Observations
    Adding, 16
    Deleting, 17
    Selecting a Subset, 18
    Sorting, 17
Odds Ratio, Logistic Regression, 130
One-Sample t Confidence Interval, 80
One-Sample t Test, 81
One-Way Analysis of Variance, 110
Opening
    Excel Data File in SPSS, 14
    SPSS Data, Existing, 9
    SPSS Output, Existing, 23
Outliers, 50

Output
    Deleting, 22
    Opening, 23
    Printing, 21
    Saving, 22
Pattern Fill, Changing on a Chart, 32
Percentages, Two-Way Tables, 93
Percentiles, 48, 51
Pie Charts
    Editing, 35
    Grouped Data, 38
    Ungrouped Data, 33
Power, 79
Prediction Intervals, Regression, 105
Printing, Data, Output/Charts, 21
Probability Distributions
    Binomial, 70
    Normal, 55, 72
    Other, 72
    t, 80
Proportions, Inference for, 90
Quartiles (see Percentiles)
Quitting SPSS, 27
r, 60, 102, 108
r^2, 61, 102, 108
Random Assignment, 68
Random Number Generator, 58
Random Sampling, 67
Recoding a Variable, 19
Reference Line on a Residual Plot, 64
Regression
    Diagnostics, 66
    Inference, 100
    Line on Scatterplot, 61
    Logistic, 128
    Multiple Linear, 106
    Simple Linear, 60, 100
Residuals, 63, 103, 104
Sampling Distributions, 73
Saving
    Excel File in a Format Usable by SPSS, 13
    SPSS Data, 7
    SPSS Output, 22
Scale of Measurement (nominal, ordinal, scale), 2
Scatteplots, 59, 104
    Adding Regression Line to, 61
Selecting a Subset of Observations (or Cases), 18, 96

Sign Test, 86
Simple Linear Regression, 60, 100
Simulations, 73, 77, 78
Slope, 60
Sorting Observations, 17
Standard Deviation, 48, 51
Stemplots, 40
String Variables, 3
Syntax, 73, 75, 77, 79
Tests
    Chi-Square, 95
    F (Analysis of Variance), 111, 119
    Kruskal-Wallis, 126
    Levene's, 89,
    Matched-Pairs t, 84
    One-Sample t, 81
    Sign, 86
    Two-Sample t, 88
    Wilcoxon Rank Sum, 121
    Wilcoxon Signed Rank, 123
Tests of Significance, 78
Time Plots, 46
Transforming Variables, 54
Two-Sample t Test, 88
Two-Way Analysis of Variance, 116
Two-Way Tables, 91
    Inference for, 91
    Percentages with, 93
Unit of Measure, Changing, 54
Variable
    Adding, 16
    Creating, 54, 55
    Deleting, 17
    Labels, 4
    Names, 3
    Recoding, 19
    Transforming, 54
    Value Labels, 5
Weight Cases, 92, 121
Width
    Changing Column, 4
    Changing a Variable, 3
Wilcoxon Rank Sum, 121
Wilcoxon Signed Rank, 123
Word, Copying From SPSS into, 24
x axis, Changing on a Chart, 43
y axis, Changing on a Chart, 45
y-intercept, 60